WEIRDEST MATHS

ABOUT THE AUTHORS

David has a PhD in astronomy from the University of Manchester. For the past thirty-five years he's been a freelance science writer and is the author of nearly fifty books on subjects such as cosmology, physics, philosophy, and mathematics. His website www.daviddarling. info, and YouTube channel, discovermaths, are widely used online resources. He also tutors students in maths and physics, which is how he first met Agnijo at the age of thirteen.

Agnijo was born in Kolkata, India, but has spent most of his life in Scotland. His extraordinary mathematical talents were recognised at an early age. In 2018 he came joint first in the International Mathematical Olympiad, recording a perfect score and affirming his status as one of the world's most outstanding young mathematicians. He is now continuing his studies at Trinity College, Cambridge.

WEIRDEST MATHS

At the Frontiers of Reason

DAVID DARLING
AND
AGNIJO BANERJEE

ONEWORLD

A Oneworld Book

First published by Oneworld Publications in 2020

Copyright © David Darling and Agnijo Banerjee 2020

ISBN 978-1-78607-805-6
eISBN 978-1-78607-806-3

Typeset by Tetragon, London
Printed and bound in Great Britain by Clays Ltd, Elcograf S.p.A.

Oneworld Publications
10 Bloomsbury Street
London WC1B 3SR
England

Stay up to date with the latest books,
special offers, and exclusive content from
Oneworld with our newsletter

Sign up on our website
oneworld-publications.com

MIX
Paper from
responsible sources
FSC
www.fsc.org   FSC® C018072

The difference between the poet and the mathematician is that the poet tries to get his head into the heavens while the mathematician tries to get the heavens into his head.

– G. K. Chesterton

Mathematical truth is immutable; it lies outside physical reality... This is our belief; this is our core motivating force.

– Joel Spencer

If I had a dollar for every time algebra has helped me, I'd have x dollars.

– Anonymous

Contents

Introduction

WE BEGAN WRITING our first book together when one of us was a 61-year-old science writer and the other a 15-year-old schoolboy. It seemed an unlikely combination but Agnijo was no ordinary teenager. His father, who'd heard that I (David) do some tutoring in maths and science, asked if I'd be available to work with Agnijo to help develop his wider knowledge of maths, modern physics, and areas where science and philosophy overlap. His school was running out of things to teach him and he was taking exams four years ahead of his age group (nearly always scoring full marks).

From the outset, it was clear that Agnijo had outstanding abilities. He could perform extraordinary feats of mental arithmetic and had a phenomenal memory. The first time he came to me I lent him a copy of a book I'd written about ten years before called *The Universal Encyclopedia of Mathematics*, a fairly dense 400-page A-to-Z of everything under the mathematical sun. A week later he brought the book back – not only had he read it and memorised large chunks but he'd also found several minor mistakes! From then on, our sessions weren't so much conventional tutorials as they were conversations at a graduate level, on everything from the nature of dark energy to ways of naming numbers vast beyond imagination.

At some point in 2015 I suggested to Agnijo that we write a book together, dividing up the chapters according to our special interests and cross-checking each other's work. Fortunately, my publisher, Oneworld, saw the merit in our unusual partnership. Our proposal was accepted and the start of 2018 saw the fruits of our collaboration – *Weird Maths* – appear in print.

If any seventeen-year-old mathematicians have had a bigger year than Agnijo did in 2018, they must be few and far between. Shortly after becoming a published author, Agnijo was selected for the UK team at the prestigious Balkan Mathematical Olympiad and then to represent Britain at the 2018 International Mathematical Olympiad. His one and only appearance at an IMO saw him take joint first place with a perfect score – the best result by a UK competitor in twenty-four years. No sooner had he returned to Britain than he was on his travels again, this time to India with his family, to help promote a new edition of *Weird Maths* there. As if that were not enough, he had only a few weeks to recover after coming back to British soil before starting his maths degree at Trinity College, Cambridge, the alma mater of so many of his mathematical heroes, including Srinivasa Ramanujan, G. H. Hardy, and Charles Babbage.

In *Weirdest Maths*, the final book of the *Weird Maths* trilogy, we trek far and wide across the landscape of this strangest and most wonderful subject. We explore the maths of sport, life, and the universe as a whole. We ask what makes a mathematical genius and follow the adventures of Agnijo from elementary school prodigy to first place in the world's greatest maths competition. We look at maths in fiction, examine the claim that beauty is a reliable guide to truth, and ask what mathematicians may accomplish over the decades ahead.

To many people maths may seem a dull, hard subject – one of the necessary evils of school to be avoided later on as much as possible. But maths is infinitely greater than a system of calculation. It permeates everything around and within us – music, art, the natural world, and the human mind – forming an invisible infrastructure of reality. Mathematics, far from being dry and difficult, is vibrant and fascinating, as wonderful as it is weird.

CHAPTER 1

Genius

Talent is a flame. Genius is a fire.

— Bernard Williams

WHEN HE WAS six, John von Neumann could multiply and divide two 8-digit numbers in his head. A couple of years later he'd progressed to solving tough problems in differential and integral calculus. He'd also amuse his parents' friends by chatting in ancient Greek or reciting whole pages of a telephone directory that he'd memorised at a glance – early signs of the startling ability that would blossom in adulthood. Economist Paul Samuelson said of von Neumann that he had 'the fastest mind' of anyone he'd met. Polish-born British mathematician Jacob Bronowski, in the 1973 documentary series *The Ascent of Man*, considered him 'the cleverest man I ever knew, without exception' (the second cleverest he judged to be Italian-American physicist Enrico Fermi).

These days 'genius' is a much overused label. It's also, like beauty, an imprecise one that's partly in the eye of the beholder. The claim to be able to measure the threshold of genius with a single number – an 'intelligence quotient' or IQ – just doesn't bear scrutiny. A commonly applied type of IQ test, based on the work of French psychologist Alfred Binet in the early twentieth century, puts a figure of 100 or thereabouts on average intelligence and rates genius as anything above about 160. But like the cryptic crossword puzzles in a newspaper, IQ tests are things you get better at with practice and age (up to a point). They also favour certain types of thinker. Who'd feel happy about ranking the relative genius of Beethoven, Picasso, and Einstein based, say, on their scores in the entrance test to Mensa? Richard Feynman, by common consent one of the brightest theoretical physicists of the twentieth century, and co-winner of the 1965 Nobel Prize for Physics (with Julian Schwinger and Sin-Itiro Tomonaga), managed just 125 in a high school IQ test. My co-author, Agnijo, took the Mensa entrance test at age twelve and achieved a maximum possible score of 162, putting him ahead of Stephen Hawking at the same age. But he's modest enough to recognise that such comparisons are meaningless unless a great intellect is turned to good use later in life.

What is genius – and, in particular, *mathematical* genius? Does the potential for it lie within each of us if only we knew how to tap it? Or does the spark of genius have to be there from the start, in the makeup of an individual's brain? There are no easy answers because the forms in which exceptional ability and achievements come are so varied.

John von Neumann, for instance, though unquestionably a genius by any definition, had a privileged upbringing. Born to Jewish parents in Budapest in 1903, he was given every advantage money could buy including, at the age of eight, entrance to the Fasori Evangélikus Gimnázium, one of three prep schools at the pinnacle of a superb education system in the Hungarian capital – for those who could afford it. Between the late 1890s and the 1930s, this elite system turned out a

John von Neumann, widely regarded as one of the greatest mathematical geniuses of the twentieth century, shown here in 1956.

generation of mega-intellects who played important roles on the world stage of science and maths: von Neumann himself, mathematician and space engineer Theodore von Kármán, radiochemist George de Hevesy, physicists Leó Szilárd, Eugene Wigner, and Edward Teller, and extraordinarily prolific mathematician Paul Erdős, among them. Most of these talented Jewish Hungarians ended up in the United States in the first half of the twentieth century and soon gained a reputation for their almost superhuman abilities. On one occasion, Szilárd was asked why no intelligent extraterrestrials had been found despite the seeming likelihood that they existed – the so-called Fermi Paradox. He replied: 'They are already here among us – they just call themselves Hungarians.'

No one would claim that eastern Europeans are inherently smarter than anyone else. All the Hungarian example shows is that the right upbringing and education can help foster intellectual attainment, but there's surely more to genius than that. Some of the greatest mathematicians the world has ever known came from humble backgrounds.

Take Carl Friedrich Gauss, born in 1777 in what's now Lower Saxony, Germany. Today, he stands shoulder to shoulder with the likes of Euclid, Isaac Newton, and Leonhard Euler as a colossus of mathematics. Yet his origins were humble. His father turned his hand to gardening, bricklaying, butchering, and anything else that would help make ends meet. His mother could neither read nor write and never recorded the date of Carl's birth. She did remember, though, that it was on a Wednesday, eight days before the Feast of Ascension, which, in turn, is the fortieth day of Easter. In time, Gauss came up with a formula that gave not only his own birthdate but also the date of Easter in any year, past or future.

Even as a toddler, Carl's talent with numbers stood out: he could do sums when he could barely speak. Just after he'd turned three, he spotted a mistake in his father's tax calculations. As a seven-year-old, he solved a problem in seconds that his teacher had expected would keep his class busy for ages. The problem was to add together all the numbers from one to a hundred. Gauss quickly spotted that the sum could be broken down into fifty pairs – $(1 + 100) + (2 + 99) + \ldots + (50 + 51)$ – each of which added up to 101, giving a total of 101 times 50, or 5050. By the age of ten he'd discovered an important result in maths called the binomial theorem that, unbeknown to him, had been derived earlier by Newton. Word spread of the young prodigy's achievements and he found a sponsor in the Duke of Brunswick who offered to fund his further education.

Thanks to his benefactor, Gauss was able to attend the Collegium Carolinum – where he earned a degree in maths at the age of eighteen – and then move on to the prestigious University of Göttingen for his graduate studies. In 1796, a year after receiving his doctorate, he cracked a major problem in geometry by showing that a regular polygon with 17 sides could be constructed using just an unmarked ruler and a compass. The Greeks had known how to construct polygons with 3, 5, and 15 sides with ruler and compass alone but making the heptadecagon in the same way had resisted all their efforts. It was this breakthrough that persuaded Gauss to focus on maths instead of languages, at which he also excelled. Later in the same year, he discovered that every number is the sum of at most three triangular numbers (numbers of the form 1 + 2 + ... + n, for example 1, 3, 6, or 5050).

One of Gauss's most spectacular achievements was to track down a heavenly body that had gone missing. In 1801, Italian astronomer Giuseppe Piazzi discovered a faint object, which he called Ceres, that didn't appear in catalogues of known stars at the time. Piazzi followed Ceres over a period of several weeks and found that it wasn't a star at all but something in orbit around the Sun. Then he fell ill and lost track of the newcomer. Fortunately, Gauss was able to figure out its orbit, along with its whereabouts, using just the handful of observations that Piazzi had already made. As we now know, Ceres is the largest object in the asteroid belt, so large, in fact, that it's been reclassified as a dwarf planet.

Gauss showed an exceptional talent for maths even before he started school. His genius, it seems, was innate – although who knows if it would have flourished later on had not his ability been recognised and allowed to develop. In some ways his story parallels that of another genius who lived more recently.

The most extraordinary mathematician of the twentieth century was born, like Gauss, to working-class parents and had, early on, a very modest education. Yet by the age of eleven, when he encountered formal maths in school for the first time, it was already clear that Srinivasa Ramanujan was operating on a different plane. In his early teens he tutored other pupils, mastered new concepts with ease, and won a string of academic awards. In 1903, as a sixteen-year-old, he got hold of a library copy of a book with the disarmingly simple title *A Synopsis of Elementary Results*, which, in fact, was a dense collection of about five thousand results in pure maths based on the notoriously challenging Mathematical Tripos at Cambridge. Not satisfied merely to absorb the book's contents, Ramanujan set out to derive all of its results himself with no outside help. In the process he came up with a wealth of other extraordinary conclusions that seemed to spring from nowhere.

This almost manic creativity, with no obvious point of origin, became a hallmark of Ramanujan's work. To the end of his life, he attributed all of his major insights and discoveries to a singular source beyond logic – the Goddess of Namagiri (his hometown) who, he said, appeared to him in visions and revealed formulae, which, upon waking, he'd seek to verify. Ramanujan's proofs, however, were often incomplete, making it hard to check them or sometimes even to make sense of his propositions. They were also, sometimes, just plain wrong.

It's possible that Ramanujan would have remained in relative obscurity had he not, in his twenties, written a series of letters to distinguished British mathematicians. Only one of them took him seriously. Fortunately, that one happened to be G. H. Hardy, famed Cambridge scholar and distinguished number theorist who had himself been precocious as a child. While still a toddler he wrote down numbers into the millions and, later, when taken to church on the Sabbath he'd pass the time factorising the numbers of hymns. In Ramanujan's writings, Hardy recognised something very special indeed. Some of the Indian's results corresponded to known maths, but of a very advanced nature and arrived at by unfamiliar means. Other results seemed utterly new but, in Hardy's opinion, probably true 'because, if they were not true, no one would have the imagination to invent them'. In Ramanujan's obituary, which, sadly, Hardy would pen just seven years later, he wrote that Ramanujan was 'a mathematician of the highest quality, a man of altogether exceptional originality and power'. On his personal scale of maths ability, Hardy scored himself a modest 25, another close colleague at Cambridge, John Littlewood, 30, and David Hilbert, the most renowned mathematician of the time, 80. Ramanujan he rated at 100.

Hardy invited Ramanujan to join him at Cambridge and for a few years the two formed a formidable team, at the very college, Trinity, where Agnijo now studies. Hardy taught the younger man how to set down proofs in an orthodox way so that they could be published in academic journals and checked by other mathematicians. At the same time, he was aware that it was neither possible nor desirable to give the Indian a conventional education in all the areas of maths that he'd missed. Hardy understood well the danger of such an education: that it can stifle the kind of extreme creativity that's so often the sign and greatest product of true genius.

Knowing too much about a subject can make us overly

cautious. Having a lot of conventional wisdom may make us doubt our own hunches and intuition because we're more likely to think that any seemingly good ideas that pop into our heads are wrong if they don't square with what we've previously learned. Had Ramanujan received an expensive but traditional education, would his genius have burned so brightly and uniquely? For sure, mathematical genius needs some foundation on which to build, but what's the optimal amount of formal schooling to nurture genius but not at the same time crush it with conformity?

The strangest aspect of Ramanujan's genius was his conviction that it had a supernatural source. Artists and musicians have often, especially in the past, expressed the view that their work was divinely inspired. Kahlil Gibran, the Lebanese-American mystic poet and artist, wondered: 'Am I a harp that the hand of the almighty may touch me or a flute that his breath may pass through me?' It's more unusual for mathematicians or scientists to regard themselves as agents of a higher power, but it's common for them to talk about the importance of sudden inspiration. Famously, the German chemist August Kekulé ascribed his discovery of the benzene ring to a dream in which he saw a serpent with a tail in its mouth – an ancient symbol, known as the *ouroboros* (Greek for 'tail-devourer'), representing an endless cycle or loop. French philosopher and mathematician René Descartes and mathematician Henri Poincaré also accounted for some of their important discoveries in terms of pictorial revelations from the unconscious. In a chapter titled 'Mathematical Creation' from his 1904 book *The Foundations of Science*, Poincaré put it this way:

> The subliminal self is in no way inferior to the conscious self; it is not purely automatic; it is capable

of discernment; it has tact, delicacy; it knows how
to choose, to divine.

Poincaré is another of those widely regarded as being among
the greatest of mathematical geniuses. He's been described as
'the last universalist' because after his time (he died in 1912)
maths spread so far and wide that even the most penetrating
minds on Earth could no longer master all aspects of it.

Swiss psychiatrist and psychoanalyst Carl Jung spoke of
the ancient tradition of knowledge being imparted through
dreams and visions and how twentieth-century rationality
tended to devalue its significance. Yet Albert Einstein, the
towering figure in physics from the past hundred years, was
quick to acknowledge that his biggest breakthroughs came
out of the blue. He recalled that, one night in the spring
of 1905, 'a storm had broken loose in my mind' and in the
morning it was as if he had the master plan for the universe
in his grasp. The days and weeks that followed saw Einstein
working feverishly, non-stop, filling thirty-one pages of notes
that formed the basis for his special theory of relativity – a
new physics of space and time.

When insights come in this way, in dreams or daydreams,
or in waking from a deep sleep, it isn't hard to see how they
might be attributed to a deity or some mysterious cosmic
influence. Ramanujan was steeped in the religion of his family
and birthplace, which habitually gave credit for happy outcomes to Namagiri Thayar, the local form of Lakshmi, the
Hindu goddess of wealth and success. It would be natural for
visions of her, constructed in his mind, to overlap with subconscious mathematical musings – his two great obsessions
fused into one. Even Einstein spoke sometimes of his ideas
coming from God, although he professed to believe not in

a personal god but instead something more along the lines of Spinoza's god – 'a superior mind that reveals itself in the world of experience'.

We're all familiar with occasions when we go to bed with some problem niggling away at us only to wake to find the solution there crystal clear, arrived at effortlessly sometime in the night. Our brains are just naturally good at sorting things out without conscious intervention – in fact, in many cases, thinking just seems to get in the way. But no amount of R&R can help if we're trying to do something difficult without first having the necessary knowledge or skill. Great mathematicians may have great insights but they already know a lot about their subject and spend much of their waking time involved with it. The same is true in every walk of life. A top tennis player will speak of being 'in the zone', when shots flow effortlessly with fluid grace and the player is effectively on autopilot. But that experience can only come after years of practice and dedication.

Argentine-American mathematician and computer scientist Gregory Chaitin, who's made important contributions to information theory, has described the intellectual equivalent of being in the zone:

> I can only look at my own experience creating a new mathematical theory and say I don't know where it comes from… I seem to be in some kind of energised or more perceptive state and it's a wonderful state to be in. It doesn't last long. It feels wonderful.

Whenever the subject of extreme ability crops up, whether it's intelligence, artistic skill of some kind, or athletic prowess, there's the age-old issue of nature versus nurture. Both, on their own, have limits. With the best will – and training regime – in the world, the reader, even in his or her prime, would never be able to run as fast as Usain Bolt. Nurturing will only get you so far if the basic physical ingredients aren't in place. At the same time, a potential genius might go to waste without the right encouragement and support. In some cases, it seems, nature takes the lead and nurture follows, as in the case of Gauss and Ramanujan. In other instances, exceptional mathematicians and scientists have emerged after not-very-promising starts to their careers.

Jacques Hadamard was a French mathematician who rose to fame at the end of the nineteenth century with his proof (arrived at independently by Belgian contemporary Charles de la Vallée Poussin) of the prime number theorem. This theorem has to do with how prime numbers are distributed on the number line and bears upon the Riemann hypothesis – the biggest unsolved conundrum in maths. Yet, as Hadamard pointed out, he was a late bloomer: 'in arithmetic, until the fifth grade [ten or eleven years old], I was last or nearly last.'

The same was true of Hermann Grassmann, one of the founders of linear algebra – a subject that now has all kinds of applications in science, from quantum mechanics to machine learning. A couple of centuries ago, the young Grassmann was studying at Stettin Gymnasium, in what was then Prussia, where his father taught maths and physics. Grassmann senior wasn't optimistic about his offspring's mathematical potential and recommended that perhaps he become a gardener instead. Luckily for the world, Hermann persevered with his academic studies, although at university, in Berlin, he took courses not in maths but theology, classical languages, and philosophy. He came back to maths in a modest way, taking a year out to prep himself in the basics before sitting exams so that, like his dad, he could teach the subject in secondary schools. It was while writing an essay for one of these exams, on the theory of tides, that Grassmann introduced an entirely new mathematical approach – what became known as linear algebra – together with the concept of vector spaces.

Grassmann got little or no credit for his innovations at the time. He qualified to become a schoolteacher and, in 1852, at the age of forty-three, rose to his late father's position at the Stettin Gymnasium with the title of 'professor'. But his greatest ambition, to teach at a university, was stymied by a report to the Prussian Ministry of Education by another mathematician, Ernst Kummer, in which he described Grassmann's ground-breaking essay on tides as 'commendably good material expressed in a deficient form'.

Such criticism became a recurrent theme in Grassmann's life. Over and over again, with a handful of exceptions, his peers attacked his methods and failed to grasp the underlying importance of his ideas. His first great published work, *Ausdehnungslehre* ('Theory of Extension'), in 1844, was almost totally ignored, as was a new version of the book eighteen years later. Eminent mathematicians such as August Ferdinand Möbius, Augustin-Louis Cauchy, and Giuseppe Peano were aware of what Grassmann was doing, and Peano was generous enough to acknowledge the part that Grassmann's concepts played in his own treatment on the foundations of natural numbers. But the fact is that Grassmann was too far ahead of his time to be properly appreciated. He also lacked the language and mathematical tools, such as those of set theory, which hadn't yet been developed, to be able to express his ideas rigorously. In retrospect, he's seen as one of the few people ever to invent a whole new branch of mathematics pretty much single-handed. But it wasn't until the first quarter of the twentieth century, when Hermann Weyl and others came up with formal definitions of the key concepts at the heart of linear algebra, notably that of a vector space, that Grassmann's genius and the extent of his achievement were recognised.

Who knows how many geniuses in the past lived at a time before their ideas could be properly appreciated and take root. At least Grassmann won posthumous credit for his daring foresight. Scores of others with the capacity to revolutionise maths or science must have gone unnoticed over the centuries simply because they were born in the wrong time or place.

Albert Einstein, it happens, got lucky on both counts. Physics was ripe for a revolution and in Europe theoretical developments were taking place that prepared the ground for it. It needed only a genius, armed with deep enough insight and a rebellious nature, to usher in the new scientific paradigm.

Ask someone to name a genius and the chances are they'll say 'Einstein' and have in mind the image of an old man with smiling eyes and long, wild hair. Einstein's adult years are well documented, his childhood less so. It's sometimes said that he was a late developer, that he couldn't speak well until he was about nine, and that his teachers thought he might have learning difficulties. What seems more likely, as some psychiatrists have suggested, is that Einstein had Asperger's syndrome, the traits of which include a preoccupation with narrow, often arcane interests, disregard for social mores, lack of interest in general conversation, and sometimes unkempt appearance. The diagnosis is bound to be uncertain and controversial because Einstein is no longer with us and the condition wasn't fully recognised until after his death. But certainly, as a child, Einstein was intensely focused on maths. Aged twelve he was given a geometry textbook by a friend of the family, Max Talmey, a medical student at the time, who mentored the young Albert on weekly visits to his home. Talmey later recalled that, in the space of one summer: '[He] had worked through the whole book. He thereupon devoted himself to higher mathematics... Soon the flight of his mathematical genius was so high I could not follow.' In the same year, Einstein started teaching himself calculus, a subject he mastered within a year or two. Over the same period, he absorbed Immanuel Kant's *Critique of Pure Reason*, a dense, difficult treatise that would baffle most adults. In other school subjects, Einstein was no more than average, which is why he didn't gain entrance to the Federal Institute of Technology in Zurich (ETH) at his first attempt. As well as being narrowly focused, he was also, as fits an Asperger's profile, socially remote. In his own words:

> I am truly a lone traveller and have never belonged to my country, my home, my friends, or even my immediate family, with my whole heart; in the face of all these ties, I have never lost a sense of distance and a need for solitude...

When he was finally accepted at ETH, after earning a diploma from a minor college in Switzerland, Einstein didn't always impress those who taught him. One of his lecturers, Hermann Minkowski, called him a 'lazy dog' who 'never bothered about mathematics at all'. Of course, that wasn't really true. Einstein cared deeply about maths and physics – just not always those aspects or problems presented to him as part of his formal education.

After he graduated, Einstein, like many independent thinkers, found it hard to find a job but after a couple of years was hired as a clerk in a Swiss patent office thanks to the father of one of his academic friends. The work wasn't too demanding so it left Einstein with plenty of time to develop his own ideas. In 1905, his *annus mirabilis*, he published a series of papers, on the photoelectric effect, Brownian motion, the special theory of relativity, and the equivalence of mass and energy, any one of which might have won him a Nobel Prize (although, in fact, only the first did). Einstein was now twenty-six and at the peak of his powers. He'd remain at that summit for perhaps another decade during which he hatched a radically new theory of gravity – the general theory of relativity. But after 1915 his creativity fell away and he pioneered no more ground-breaking science for the rest of his life.

In his 1940 memoir, *A Mathematician's Apology*, G. H. Hardy wrote: 'No mathematician should ever allow himself to forget that mathematics, more than any other art or science, is a young man's game.' Hardy's sentiment is commonly extended to include physics, especially theoretical physics, which is highly mathematical. Certainly there are many examples to bolster the argument – Einstein being a case in point. His genius burned intensely for a dozen or so years after the turn of the century, but then flickered out.

By the time Einstein arrived at his final academic home, the Institute for Advanced Study (IAS), in Princeton, New Jersey, in 1933, he'd begun a long and futile quest for a unified theory of gravity and electromagnetism. At the IAS he found himself in the company of two other intellectual giants, both unusual in their own ways: the Austrian Kurt Gödel, a logician, who became Einstein's closest friend, and John von Neumann.

As premier geniuses of the twentieth century, Einstein and von Neumann make an interesting contrast. Einstein today is overwhelmingly the better known, but von Neumann's achievements span a broader range and began earlier in his career. By the age of nineteen, von Neumann had published two major mathematical papers, the second of which gave the modern definition of so-called ordinal numbers – numbers that can be used to generalise the concept of natural numbers. He later pioneered game theory and early electronic computers and played a prominent role in the Manhattan Project, the top-secret US programme to develop an atomic bomb.

One of von Neumann's colleagues on the Manhattan Project was fellow high-IQ Hungarian Eugene Wigner. The two had been a year apart in the same elite school in Budapest. When asked why the Hungary of his generation had spawned so many geniuses, Wigner, who won the Nobel Prize in Physics in 1963, replied that von Neumann was the only genius. Perhaps being a close friend from childhood, Wigner was biased but he said of von Neumann 'only he was fully awake'. Comparing him with Einstein, however, he commented:

> Einstein's understanding was deeper even than von
> Neumann's. His mind was both more penetrating
> and more original than von Neumann's.

Various factors, it seems, contribute to what we call genius and the forms it may take: speed of thought (at which von Neumann, by all accounts, was exceptional), depth of understanding (at which, according to Wigner, Einstein excelled), originality, creativity, and so forth. Sometimes, too, genius may be narrow in its focus – as in the case of Einstein or Ramanujan – while at other times, as illustrated by von Neumann, and to an even greater extent by some Renaissance figures such as Leonardo da Vinci, it can range over many subjects. For all his recognition during life, von Neumann doesn't have the celebrity status today of Einstein whose office at the IAS was just down the hall. Yet while Einstein essentially stagnated after his arrival at the Institute, von Neumann continued to flourish, taking on one massively difficult challenge after another right up until the end of his relatively short life. From the maths of quantum mechanics he'd pivot to practical problems in weather prediction or hydrology, the foundations of computing, or cellular automata. He had an outstanding mastery of many branches of mathematics, which he could bring to bear in his work in physics and computation. Also, he had an astonishing, apparently photographic memory. The mathematician and computer scientist Herman Goldstine, a fellow collaborator on the ENIAC computer project, wrote:

> As far as I could tell, von Neumann was able on once reading a book or article to quote it back verbatim; moreover, he could do it years later without hesitation... On one occasion I tested his ability by asking him to tell me how *A Tale of Two Cities* started. Whereupon, without any pause, he immediately began to recite the first chapter and continued until asked to stop after about ten or fifteen minutes.

As I'm writing this, it brings to mind a connection between these geniuses of the recent past, Einstein and von Neumann, and my co-author, Agnijo. The link is in the form of another outstanding thinker – the English-later-American physicist and mathematician Freeman Dyson. In the year I was born, 1953, Dyson was given a permanent post at the IAS (having previously worked there temporarily in the late 1940s) and an office close to those of both Einstein and von Neumann. Of von Neumann he wrote:

> Johnny's unique gift as a mathematician was to transform problems in all areas of mathematics into problems of logic. He was able to see intuitively the logical essence of problems and then to use the simple rules of logic to solve the problems.

Dyson, who died in 2020, was himself extraordinarily creative throughout his long career, starting at the age of five by showing a fascination for large numbers and calculating how many atoms are in the Sun. He studied under G. H. Hardy at Cambridge and gained a reputation for going out of his way to counter orthodoxy. A friend of his, neurologist and author Oliver Sachs, said: 'A favourite word of Freeman's about doing science and being creative is the word "subversive".' In 1947, he moved to the United States where he quickly formed an association with Richard Feynman at Cornell University. A couple of years later he stamped his mark on the world of science by proving the equivalence of two different formulations of quantum electrodynamics: Feynman diagrams and the so-called operator method developed by Schwinger and Tomonaga.

For four decades, from 1953 until his retirement in 1994 (he remained emeritus), Dyson was one of the leading lights at the IAS, bringing his genius to bear on problems as diverse as nuclear space propulsion, detecting alien civilisations, and uncovering a possible link between the distribution of prime numbers and energy levels in the nuclei of heavy elements. In 2004 he was kind enough to read a copy of my book *The Universal Book of Mathematics* and point out a couple of minor errors, which no one up to that point had noticed. When Agnijo started coming to me for tuition in 2013, just after he'd turned thirteen, I loaned him a copy of the book and invited him to keep it as long as he liked. He brought it back the following week, having read all 400 pages, along with a short list of errata – including the two that Dyson had spotted!

It's too early to say what mark Agnijo will leave on the world. Now studying at Trinity College, Cambridge, alumni of which include Charles Babbage, James Clerk Maxwell, Niels Bohr, Bertrand Russell and, as we've seen, Hardy, Ramanujan, and Dyson, having shared first place in the 2018 International Mathematical Olympiad, he has extraordinary powers of problem-solving and memory. Moreover, his talents are innate. Up to the age of seventeen he attended ordinary

Agnijo at the age of twelve.

state schools in Scotland and his outstanding maths ability is almost entirely self-developed, though, of course, he's been supported and encouraged by his family and others around him.

I've asked Agnijo about his genius – because there is no doubt that's what it is. Where does it come from? What makes him different from almost everyone else in this particular field – mathematics? He has no ready answers. It isn't easy for a person to say why they're exceptional when, to them, exceptional is the norm. He points to his deep passion for the subject. But he freely admits there's probably something unusual about his brain because, like many outstandingly clever people, his talents became clear to others at a very early age, before he'd been exposed to anything but simple arithmetic in school. He also has a fantastic memory, which helps him do well in most academic subjects without the effort most of us would need.

Perhaps a feedback effect goes part of the way to explaining genius. A person may be born with a brain better equipped to do maths just as others start out with an advantage, sometimes genetic, when it comes to, say, athletics or singing. If we're naturally good at something we'll tend to do it more often and get even better at it because it comes easily and is enjoyable. Then others will praise us and we'll be encouraged to develop our talent more and more, perhaps to the point of obsession, until it becomes truly exceptional.

Research by psychologists suggests that certain other skills often go hand in hand with maths wizardry. For instance, children who, early on, show an aptitude for maths also tend to be good at solving visuospatial problems. There's evidence, too, that being good at maths is tied to a more general capacity to spot hidden structures in data. This could explain why it's common to find people who excel at both maths and music, and why training at chess can help improve maths scores – both music and chess have complex data structures at their heart. Einstein famously claimed that images, feelings, and musical structures formed the basis of his reasoning rather than logical symbols or mathematical equations. But the fact remains, we don't fully understand why some people develop into masterminds of maths. And so complex is the problem that it may take more than a touch of genius to figure out.

While co-authoring this book with Agnijo, I asked him where he thought his spectacular maths ability came from. He said:

> I don't know exactly. I think 'genius' – whatever that means – always involves passion. Passion leads you to get very deeply involved with something and become exceptionally good at it. Also, I suppose there must be something in the infant brain that predisposes some people toward genius – but what that is is still a mystery.

CHAPTER 2

Sporting Chance

Baseball is 90 percent mental and the other half is physical.

— Yogi Berra

WHAT MAKES A cricket ball swing? Is hitting a baseball really the most difficult feat in sport? How can the equations of fluid dynamics help swimmers achieve record-breaking times? Maths today permeates sport – underpinning the science behind the sports themselves and the methods athletes use to achieve the highest possible performance. It can make the difference between success and failure in events that depend on split-second timing, extreme accuracy, peak strength, and perfect preparation.

Among the earliest of competitive sports was the discus throw, records of which go back to the eighth century BCE. It was one of the events included in the ancient pentathlon, along with the stadion (a 180-metre sprint), the javelin throw, the long jump, and wrestling, held for the first time in the eighteenth Olympic Games of 708 BCE. The object was to hurl a solid bronze disc, weighing about four kilograms (nine pounds), as far as possible, the longest of five attempts being

Discobolus – a Roman copy in marble, from the second century CE, of Myron's bronze original *Discobolus* (c. 460 BCE).

taken for each athlete. The technique of rotating the body, while transferring the weight from one foot to the other, and sweeping the arm around in a rising arc to the point of release, is partly captured in the famous statue known as *Discobolus* by Athenian sculptor Myron of Eleutherae. Although the original Greek bronze is lost, many Roman copies in both marble and bronze survive.

The discus throw, as a competitive sport, was revived in the 1870s in Germany and was included in the first of the modern Olympic Games, held in Athens in 1896. Shortly after, the modern method of rotating the whole body through one and a half turns before release was introduced by Czech athlete František Janda-Suk, who studied the statue of *Discobolus* and then used the technique to win Olympic silver at the Games in 1900.

Today's discus is smaller and lighter than that used by

the ancients – 22 centimetres in diameter and two kilograms (4.4 pounds) in weight in the case of the men's discus. Lens-shaped, it tapers from a maximum thickness of 46 millimetres in the centre to 12 millimetres at the edge and is made of plastic, wood, fibreglass, carbon fibre, or metal, with a metal rim and a metal core. The competitor stands in a circle, 2.5 metres in diameter, initially facing away from the direction of the throw. In the case of a right-hander, he or she then spins anticlockwise, through one and a half rotations, to gain momentum before the release. To be a valid throw, the discus must land within a marked 34.92-degree sector. The reason for this seemingly strange choice of angle is that it's easy to measure out in terms of a triangle with sides that are in a simple ratio: the ends of the sector lines being connected by a third line that's exactly 60 percent as long.

For much of the sport's history, athletes perfected the most successful methods of throwing exclusively through a process of trial and error. Newcomers learned the art by copying the techniques of leading exponents. But in recent times, a knowledge of physics, underpinned by maths, has played a decisive role in giving athletes a competitive edge.

Understanding the aerodynamics of the discus is crucially important. In flight, the discus becomes a fast-spinning wing, its rapid rotation giving it gyroscopic stability and its shape, when thrown correctly, additional lift. Key factors to a winning throw are the speed, angle, height, and rate of spin upon release. The characteristic wind-up, involving full use of the allowed circle, fast rotation on the balls of the feet as the athlete pivots from the back of the circle to the front, and sweeping, accelerating motion of the outstretched arm, is designed to impart the greatest possible momentum to the discus at launch. The stabilising spin, typically about 400 rpm, comes from rubbing against the ends of the fingers in the milliseconds before release. As well as being strong and agile, top discus throwers need to be tall so that, at release, their outstretched arm can reach both a good distance above the ground and a high angular momentum. To be stable in flight and gain the greatest possible lift as it flies through the air, the discus has to be tilted on release at an angle of between 37 and 42 degrees – less than that and it will lose lift, greater and it will stall.

Uniquely among track and field events, the discus throw benefits from a headwind. The spinning disc actually travels further on a moderately windy day, when thrown into the wind, because although its forward velocity is slightly reduced this is more than offset by the additional lift, and consequent hang time, afforded by the faster onward rush of air. A higher density of air also assists lift, so that lower altitudes and cooler temperatures are an advantage. Researchers at the University of Texas Institute for Geophysics found that, all other factors being equal, a discus travels about 10 centimetres further when it's a chilly 0°C than on a hot summer's day at 40°C. At the same temperature a discus thrown in Rome, 36 metres above sea level, will outdistance one thrown identically in Mexico City at an altitude of 2,225 metres. But of all external factors, wind velocity is king. Thrown into a 32-kilometre-per-hour (20-mph) headwind, the University of Texas researchers found, a discus will travel about six metres further than if thrown with a tail wind of the same speed. It comes as no surprise to learn that East Germany's Jürgen Schult had a strong breeze blowing in his face when he set a world record of 74.08 metres on 6 June 1986 – a record that still stands today.

The hammer throw and shot put also involve the athlete remaining inside a marked circle while landing their projectile within a 34.92-degree sector. But the maths and physics behind these events are quite different from those of the discus. The men's hammer is, in reality, a 7.26-kilogram (16-pound) ball – slightly less than double the weight of the women's hammer – attached to the end of a steel wire 1.22 metres long. The shot is almost identical in weight to the hammer, for both men's and women's events, but is a simple spherical object. In flight, both hammer and shot are as unaerodynamic as a rock, so that the ideal release angle for maximum range – as for any projectile that offers no lift – would seem to be 45 degrees. But in practice matters aren't that simple.

There are two main reasons why shot putters shouldn't try to throw at a 45-degree-angle. Firstly, they don't release the shot at ground level but rather, being generally quite tall athletes, from a height of two metres or more. A simple geometry calculation then shows that, for a constant projection speed, the greatest range will be achieved with a release angle of less than 45 degrees. If the projection speed is 13 metres per second, for example, and the release point is 2.1 metres above the ground, the optimum angle of projection is about 42 degrees. Second, the workings of human anatomy mean that we can throw faster at lower angles, and shot putters are no exception. This results in a trade-off between angle and speed of projection in maximising the range. Someone who can throw a shot at 12 metres per second at an angle of 30 degrees may be able to achieve only about 10 metres per second if the angle of release is 60 degrees. As a further complication, the speed-versus-angle curves for maximum range vary from one athlete to another. For most elite shot putters the optimum angle of release is between 30 and 40 degrees, depending on the individual. In every case, though, achieving exactly the right angle isn't as important as striving for top speed at the moment of release.

The goal in all throwing events is to launch the object at the optimum angle with the greatest possible speed. How this speed is generated is technically the most complex part of the sport. In the case of the hammer, the speed of rotation of the ball at the end of the wire, built up through a series of spins by the athlete, is the most crucial element. So, in this event, although the athlete must be strong it's even more important

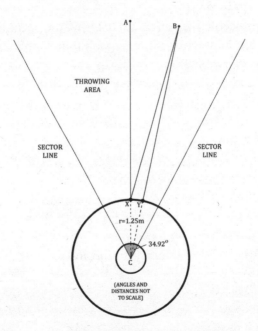

If the discus is released at X and lands at A, the recorded distance of the throw is XA, measured along the same line as the radius from C (the centre of the throwing circle) to X. If it is released at X and lands at B, however, the recorded distance is YB, again along a straight line to the centre. YB is slightly less than XA, so that some of the actual distance thrown is not credited. If, however, the discus is thrown from Y to B, there would be no loss in the distance credited for the throw.

that they're nimble on their feet in the stages before release. The shot put also involves agility in both of the main styles used – the so-called glide and spin – although more emphasis is on upper body strength and, in particular, the force with which the arm can be thrust out in the final moment.

A close look at the maths involved in all throwing events, whether discus, hammer, shot, or javelin, shows that it's better to launch straight ahead than at any other angle. This is because of something called the loss formula. The distance of the throw is measured from the point where the projectile first makes contact with the ground to the edge of the throwing circle *in a line that passes through the centre of the circle*. Launch your discus or hammer in any direction except straight ahead and a little bit of the horizontal distance travelled will be discounted because of the geometry of the situation. For example, if you throw a discus 50 metres and it lands at the extreme edge of the permitted sector – in other words, about 20 degrees away from the centre line – the distance for which you'll be credited is only 49.93 metres. While a 7-centimetre loss might not seem much, it could make the difference between a podium position and nothing in a keenly fought contest.

The maths of thrown or bowled objects takes on a different perspective in sports such as baseball and cricket. Here the goal is not to achieve great distances but to outfox the opposing batter through a ball's movement and/or speed. A baseball and a cricket ball are similar in size and weight but their construction and, in particular, their stitching is different. Laying bat on ball is a more daunting prospect in baseball because not only does the ball typically leave a professional pitcher's hand at slightly greater speeds than a top fast bowler's in cricket but also the distance to the batter is less – 16.8 metres compared with 17.4 metres, resulting in travel times of fast balls of 0.40 (baseball) and 0.45 seconds (cricket). What's more, a baseball bat is a tapering round baton, as distinct from a cricket bat with its much wider, flatter face for making contact. On the other hand, the batter in cricket has to contend with the extra complication that the ball (unless it's a 'full toss') bounces and can therefore move both through the air and off the ground in the split second before it reaches the bat.

Baseball pitchers attempt to deceive opposing batters by varying both the speed of delivery and the amount and direction of spin they apply to the ball as it leaves the fingers. A fast-spinning ball will swerve or dip through the air because of a phenomenon known as the Magnus effect, after the German physicist Heinrich Magnus, who carried out experiments on it in the nineteenth century. The spinning of a ball gives rise to a pressure difference between opposite sides of the object, which in turn generates a force at right angles to the direction of the spin. A ball spun sideways (with a vertical axis of rotation) veers either to the right or the left. One that's given forward spin, or 'top-spin' – a pitch known as a curveball in baseball parlance – will dip before it reaches the batter.

The Magnus effect doesn't play such a big role in cricket, except to some extent in the drift and dip that spin bowlers sometimes achieve. So-called 'swing', or sideways movement through the air, prized by fast and, especially, fast-medium-paced bowlers, comes about because of the asymmetry of the ball when released. One way to cause this asymmetry is to grip the ball with the fingers at an angle to the seam so that it travels through the air with more of a rough edge on one side than the other. The other, well-tried and tested method is for the bowler and fielders to rub the ball against their clothing to maintain shine on one hemisphere while allowing the other to become dull and worn. Air glides more easily over the polished surface so that the ball is effectively sucked in the direction of the rougher side.

The maths of aerodynamics is at the heart of other ball games, including golf, tennis, football, rugby, and American football. The Magnus effect arising from sideways spin explains the errant motion of a golf ball when sliced or hooked. Struck correctly, however, so as to give it a back-spin of several thousand revolutions per minute, and the ball experiences lift from this same effect, enabling it to travel higher and further. The purpose of the dimples on the surface – anywhere between 300 and 500 of them – is to induce turbulent flow, reducing the size of the low-pressure 'wake' region behind the ball and thereby reducing the net drag on the ball. At the point of contact with the club head, the force on the ball is astonishingly strong and brief, up to 18,000 newtons (4,000 pounds) lasting for half a millisecond. Struck this hard by a skilled golfer, the ball can travel over 250 metres, though if its surface were completely smooth that distance would be reduced to a mere 100 metres or so.

Mathematicians and scientists today influence every aspect of sports at the highest level, from the design of equipment to the performance of athletes. In some instances, the recipe for success can be quite specific – at least in terms of the numbers involved. In the 100 metres sprint, when rising from the blocks, the front leg knee angle should be close to 90 degrees and the rear leg knee angle approximately 120 degrees. During the first seven or eight strides body angle should increase from 45 to 60 degrees and speed to 70 percent of maximum. After 16 or 17 strides (30 metres), the body should be upright and speed at 90 percent before maxing out shortly after. Surprisingly, the maths reveals, the last 40 metres of the race should be run at under 100 percent but with increased stride pattern.

At the University of Bath, sports scientist Ken Bray has explored the maths of football and concluded that around a third of the world's best players have a strong intuitive grasp of maths, especially geometry. This gives them an edge in precision passing, taking free kicks, and positioning themselves on the field. Bray has figured out the stats for optimum free kicks. Twenty-five yards out and aimed at goal, the ball should be struck with an elevation of 16 degrees, a speed of 60 to 70 mph, and a spin rate of about 10 revolutions per second. When it comes to long-range throw-ins, researchers at Brunel University concluded that the best angle for release is around 30 degrees because most players can launch the ball fastest at this angle and so maximise their range.

Coaches use maths, or consult mathematicians, to get the best out of sportspeople and teams. In figure skating, for instance, more points are available for elements, such as spins, jumps, and flips, which are more technically difficult. But is it better for a performer to attempt a hard manoeuvre and risk botching it or go for an easier option? And what order of moves and sequence of transitions is likely to work best for them? A mathematician can analyse past performances and help put together a programme that maximises the individual's scoring potential.

Among the most complicated of scoring systems is that used in one of the most demanding of Olympic sports – the decathlon. Ten events, comprising the 100, 400, and 1,500 metres, 110 metre hurdles, long jump and high jump, pole vault, shot put, discus, and javelin, take place over two days of competition. Points are awarded for each event using a formula and a table of values determined by World Athletics – the governing body for athletics around the world. The formula for working out the number of points an athlete scores in a track event is $A \times (B - T)^C$. T is the time in seconds that the athlete achieves for the event and A, B, and C are numbers supplied by the World Athletics performance tables. In the case of the 100-metre sprint, for example, $A = 25.43$, $B = 18$, and $C = 1.81$. If you ran the race in 10.25 seconds, your score for the 100 metres would $25.43 \times (18 - 10.25)^{1.81}$, which works out to be 1,035 points. The formula for field events is the same except that the subtraction is reversed because longer distances D equate to higher points, whereas in track events, shorter times give higher values. In the case of the shot put, the formula is $A \times (D - B)^C$, and $A = 51.39$, $B = 1.5$, and $C = 1.05$. Were you to throw the shot 18.3 metres, say, you'd score $51.39 \times (18.3 - 1.5)^{1.05} = 994$ points.

When the modern system of scoring the decathlon began in 1912, to coincide with the sport's first appearance at the Olympic Games, values for A, B, C, and D for each of the events were chosen so that a world record in any event would yield a score of about 1,000 points. But as performances have improved over the years, the values provided in the standard table have been gradually adjusted. The current world record stands at 9,126 points, scored by French athlete Kevin Mayer in September 2018. If anyone were to equal the world record for every single event in the competition, they'd amass 12,500 points, while the ten best performances ever achieved by anyone during decathlons sum to a total of about 10,500.

Top decathletes are regarded as being among the greatest of sportsmen because of their all-round prowess in speed, endurance, strength, and technique. But the nature of the events and the way they're scored offer more than just a physical challenge. To be among the best in this sport means picking and choosing where to focus your training for optimum payoff. When averages are figured out for the top 100 best-ever decathlon performances, the highest scores for individual events are found to be in the long jump, 110-metre hurdles, 100 metres, and pole vault. By far the lowest average score is in the 1,500 metres, followed by the discus, javelin, and shot put, all more or less on an equal level. Connecting the most points-winning events is one common ability: flat-out sprinting speed. So, if you aspire to be an Olympic decathlete, do the maths. First make sure you're an outstanding sprinter, then build up your strength and technique for the throwing events and do some general distance running.

If decathlon scoring seems a bit dense on the theory side, it's nothing compared to a fiendishly complex system used in limited overs cricket. Probably no sport under the sun is more influenced by the weather than this most bewildering – to the uninitiated – of bat-and-ball games. Rain and 'bad light' so often lead to breaks in play that rules have had to be introduced to help decide the outcome when it's impossible for a match to run its natural course. In one-day games, where each side is allocated a fixed number of overs (sets of six-ball deliveries), the scheme that's been internationally accepted is the Duckworth–Lewis method, named after English statisticians Frank Duckworth and Tony Lewis.

Having worked as a mathematical scientist for the nuclear power industry prior to retirement, Frank Duckworth became a consultant statistician to the International Cricket Council. Among his other claims to fame are developing a way of estimating personal risk perception (the Duckworth scale) and, in his early twenties, being a tenant of John Lennon's aunt. Tony Lewis taught quantitative research methods at Oxford Brookes University. Prior to that he was a lecturer at the University of the West of England where the Duckworth–Lewis method had its genesis in an undergraduate final-year project. In 2014, data scientist Steven Stern, at Bond University in Queensland, Australia, took over as custodian and updater of the technique so that it's now formally known as the Duckworth–Lewis–Stern, or DLS, method.

There had been earlier approaches to figuring out revised targets for the batting side if the number of overs available had to be reduced because of time lost to the weather. None of these worked particularly well but the final straw came in a crucial match of the 1992 Cricket World Cup between England and South Africa. A shower of rain had stopped play for twelve minutes near the end of the game with South Africa needing to score 22 runs from 13 balls – a perfectly manageable goal with two strong batters, Brian McMillan and Jonty Rhodes, at the crease. In the brief interval the players were off the field, however, the calculation method then used, called the Most Productive Overs method, gave a revised target for South Africa of 21 runs off one ball. Frank Duckworth said: 'I recall hearing Christopher Martin-Jenkins [the BBC commentator] on radio saying "surely someone, somewhere could come up with something better" and I soon realised that it was a mathematical problem that required a mathematical solution.'

The solution that he and Tony Lewis figured out would take a while to explain in detail and needs a computer to work out accurately in real time. The key point about it is that it takes into account the 'resources' available to each team at an equivalent point in their innings. These resources are the number of overs still in hand and the number of batsmen left who could score the target number of runs. If Team 2 (the side batting second) has fewer run-scoring resources than had Team 1, their target, after a stoppage of play, is adjusted downwards according to the ratio of resources available to the two sides. On the other hand, if Team 1's innings had been interrupted, it may be that Team 2 has more resources left at the same point in their innings, in which case Duckworth–Lewis adjusts Team 2's target upwards based on the runs that would be expected to be scored on average from the extra resources at their disposal. The method cleverly takes into account experience of past one-day internationals. Records show, for instance, that the rate at which resources go down isn't uniform across the overs, but varies depending on the scoring pattern. Also, losing overs due to stoppages later in an innings usually impacts a team more than losing them early on as there's less opportunity to recalibrate the targets and, on top of this, later overs tend to be more productive. Fortunately, umpires aren't expected to solve the DLS formulae themselves: even a mathematician armed only with a calculator would struggle to get through all the complex calculations before even more overs were wasted! Instead, at all international games a computer is on hand, loaded up with the necessary software, to figure out new targets rapidly if the weather intervenes.

DLS is one of those striking examples where maths has come to the rescue in solving a tricky, practical problem. Few among the general public or sporting profession may grasp all the ins and outs of how it works, but all are happy that it *does* work.

Sport today is big business and more competitive than ever. Behind the scenes, sports scientists and mathematicians strive to come up with new designs, materials, and strategies, for higher and higher performance. Sometimes the effects can be striking and almost instantly noticeable, as happened in 2008. Suddenly, in that year, world records in numerous swimming events tumbled and by margins that fell well outside the smoothly rising curves for records of previous years. Altogether more than forty swimming world-beating times were set in a period of a few months, half of them at the 2008 Summer Olympics in Beijing. It didn't take long to figure out why. Swimmers were using a new type of suit that dramatically reduced the amount of drag on their bodies.

The new swimsuit, manufactured by Speedo, was called the LZR (pronounced 'laser') Racer or, more generically, a 'fast suit'. Developed by Italian firm Mectex in association with the Australian Institute of Sport, it was designed using data from NASA wind tunnel experiments and aerodynamic software that models virtual swimmers inside a computer. The LZR suit took to the limit a variety of factors known to increase speed through the water including hydrophobic (water-repelling) materials and fabrics that compress muscles and mould the body to make it more streamlined. Speedo's outfit even incorporated a feature borrowed from the surface anatomy of sharks – synthetic dermal denticles, or skin scales, to minimise drag. The results were spectacular. Just months after the product went on the market it was adopted by a majority of the world's top swimmers. At the Beijing Olympics, 98 percent of all swimming medals were won by racers wearing the new suit. But then came controversy. A few months after Beijing, in December 2008, at the European Short Course Championships in Croatia, seventeen more swimming world records fell. Many competitors, seeing the advantages of the LZR, especially for squeezing the body to make it slicker through the water, had taken to wearing two or more suits in layers to multiply the effect. Claims quickly followed that this amounted to 'technological doping' and officials of the sport stepped in and banned the full-body fast suit. In 2009, FINA, the international body that regulates competitive swimming, ruled that men's swimsuits could at most cover the area from the waist to the knee, and women's suits from the shoulder to the knee. They also ruled that the material used must be a textile, or woven material. Needless to say, far fewer world records were broken in the pool at the 2012 London Olympics or in any subsequent competition.

All sports to some extent are numbers games, whether in terms of the calories consumed by an athlete in training, the equations used to simulate airflow over a Formula 1 car, or the use of graph theory in the scheduling of tournaments. While not many first-rate mathematicians are also known for their sporting prowess – a notable exception being Alan Turing, who came fifth in the Olympic marathon trials for Britain – their contribution to success in everything from cycling technology to goalkeeping strategy is almost (but not quite) incalculable.

CHAPTER 3

For Your Eyes Only

Turing was a quite brilliant mathematician, most famous
for his work on breaking the German Enigma codes. It
is no exaggeration to say that, without his outstanding
contribution, the history of the Second World War could
have been very different.

– Gordon Brown

SECRET MESSAGES, LOST languages, and enigmatic codes all
appeal to our curiosity about the unknown and what's hidden
from view. They've a long history and some ciphers or codes,
and entire languages, remain a mystery to this day. At the
heart of decipherment, whether of ancient writing systems
or intentionally disguised messages, is a range of techniques
in which maths figures prominently. Today, more than ever, we
depend on the mathematics of cryptography to keep us safe and
secure in a world vulnerable to online fraud and cyber-attack.

Some dead writing systems may never be decoded because
all that's left of them are a few symbols, etched into ancient
stones or shattered tablets. The rongorongo language, traces
of which are recorded on wooden tablets found on Easter
Island, was lost to human memory by the time missionaries

arrived in the 1860s. According to the native Rapa Nui people all the wise men of the island who'd known the language had been killed by earlier invaders. Most of the remaining tablets were used for firewood so that only a handful have survived to the present day. The 120 or so main characters found on the tablets, with hundreds of variations, are mostly stylised outlines of humans and other animals, plants, or geometric shapes, written left to right and bottom to top in a manner known as reverse boustrophedon. The reader starts at the bottom left-hand corner, reads a line from left to right, then turns the tablet through 180 degrees to continue on the next line. In some cases, the information recorded is thought to be calendric or genealogical in nature, but most of it has so far proved indecipherable.

Many more specimens exist of Linear A, which, along with Cretan hieroglyphic, was one of two writing systems used by the ancient Minoans. It's been found on hundreds of tablets and signs dating back to about 2000 BCE. The existence of both Linear A and Linear B was first brought to light by British archaeologist Sir Arthur Evans in the 1890s. But whereas the latter has been figured out and the two systems share many of the same signs, Linear A has largely resisted attempts at translation. Because no one knows the ancestry of Linear A – what earlier language it may have evolved from – scholars have used similarities to Linear B to try to make sense of it. Many of the texts in which Linear A is used look as if they're lists of goods, amounts, and the names of people involved in transactions, but only a solitary word, meaning 'total', has been identified beyond reasonable doubt.

Just as intriguing is the Phaistos Disc, also Minoan and dating to the middle or late Bronze Age. This features some symbols resembling those of Linear A and B plus others that

Side A of the Phaistos Disc, on display in the Archaeological Museum of Heraklion.

are unique. A clay disc, 15 centimetres across, it bears a total of 241 tokens in spiral arrangements on its two sides, apparently made by pressing hieroglyphic seals into the clay while it was still soft. As with the cryptic Easter Island tablets, the problem facing would-be decipherers is the lack of context and of more examples of the same set of characters. A paucity of evidence has also blocked attempts to make sense of the Indus Script, a writing system found in the Indus Valley and dating to between 3500 and 1900 BCE. Although 417 distinct symbols of the Script have been catalogued, they've been found only in very short combinations, making them extremely hard to translate.

Our efforts to decipher lost human languages, even with the best mathematical tools available, suggest how hard it might be for us to make sense of extraterrestrial messages should we ever come across them. Scientists have broadcast radio signals to the stars that contain coded information about our race, our knowledge, and our world, and there are ongoing efforts to detect similar artificial signals from intelligent aliens. Presumably, smart ETs would try hard to make the content of their signals understandable, but how successful we'd be in deciphering what they were trying to say is anyone's guess. There'd be some common ground in that we were able to pick up their signals in the first place. Also, it's widely assumed that mathematics is universal, so there's the possibility, if we ever find ourselves involved in interstellar communication, of using prime numbers and other basic ingredients of maths to establish some meeting of minds.

Deciphering a message is made harder if its creator goes out of their way to cloud the meaning. And the situation's made doubly difficult if it isn't clear if what's being dealt with is a genuine message at all. Take the Voynich manuscript. Discovered by book dealer Wilfrid Voynich in 1912 in an Italian monastery, it's a medieval tome rumoured to have belonged to the Holy Roman Emperor Rudolf II of Bohemia. There's little doubt about its age – radiocarbon dating of the book's vellum puts it at mid-fifteenth century and the inks used are similarly old. But its contents are baffling. Page after page is filled with exotic depictions of medicinal plants, naked nymphs, and astrological, astronomical, or cosmological themes, together with more than 170,000 characters arranged into 35,000 or so 'words' of text, the meaning of which has eluded a century of determined effort by cryptanalysts.

The distribution of words and their structure seem to follow some internal pattern peculiar to the Voynich manuscript. For instance, certain characters occur only at the start or end of words and never in the middle of them. That

A page from the Voynich manuscript, which remains undeciphered.

pattern is foreign to any known European language. Words that differ by only one letter crop up surprisingly often and there are places where the same word appears up to three times in a row. If the materials of the manuscript had proved to be relatively recent, the whole thing would doubtless have been dismissed as a hoax. As it is, even granted its great age, it may end up suffering the same fate.

In 2004, University of Keele computer scientist and linguist Gordon Rugg delivered his verdict on the manuscript: it's pure gibberish. Rugg concluded that, although the Voynich text gives the appearance of being linguistically sophisticated, it would have been possible to generate it using cryptographic techniques that were known 450 years ago. He argued that it could have been done using a grid, known as Cardan's grille after Italian mathematician Girolamo Cardano, who invented it in 1550, together with a large table of meaningless syllables. Andreas Schinner at Johannes Kepler University in Linz, Austria, did a statistical analysis of the script and agreed that it's probably nonsensical. Yet the debate continues. Rugg's analysis has been criticised on various grounds, including the fact that the manuscript significantly predates Cardano. Whatever the truth of the matter, the Voynich manuscript is an extraordinary piece of work: either the most elaborate literary fake or the cleverest of medieval ciphers.

Much more recent is another undeciphered piece of writing. English composer Edward Elgar, famed for his *Enigma Variations*, various concertos and symphonies, and the *Pomp and Circumstance* marches, was also a keen amateur cryptographer. Some musicologists believe that within the fourteen pieces that comprise the *Enigma Variations*, Elgar managed to hide the melody of a famous song by another composer. To this mystery must be added that of the Dorabella Cipher. In 1897, Elgar wrote a cryptic pencilled note to a young female friend and admirer, Dora Penny. It was made up of 87 squiggly characters – 24 distinct ones – spread across three lines. The secret message came to light when Dora, after whom Elgar named one of his Variations, published it in her memoir, *Edward Elgar: Memories of a Variation*. At first glance, it might seem as if each different squiggle stood for a different letter of the alphabet but no analysis along those lines has come up with a credible solution. There's been speculation that it may have been in some kind of shorthand known only to Elgar and his friend (whom Elgar nicknamed Dorabella), perhaps to disguise a romantic communication between the married composer and Dora, seventeen years his junior. But Dora claimed in her memoir that she herself was in the dark

The Dorabella Cipher, written by Edward Elgar in 1897.

as to its meaning. In 2007, the Elgar Society ran a competition, offering a prize of £1,500 to anyone who could satisfactorily crack the Dorabella Cipher. However, none of the entries received, despite their ingenuity, came up with anything like a convincing translation. It's been suggested that the cipher isn't a piece of writing at all but instead a series of notes or part of a score. Given that Elgar composed his *Variations* just over a year later, it's conceivable that it's a coded snatch of the very one he devoted to his friend.

The need to send confidential information has been around as long as civilisation. Even before there were codes and ciphers, there was steganography – the art of hidden writing. All kinds of ingenious methods have been used over the ages. One of the earliest was described in 440 BCE by Herodotus in his *Histories*. A Greek called Histiaeus, according to Herodotus, shaved the head of his most trusted servant and tattooed a message to his son-in-law, Aristagoras, ruler of Miletus, on the man's bare scalp. Having waited long enough for the servant's hair to grow back he then sent him on his way with the instruction: 'When thou art come to Miletus, bid Aristagoras shave thy head, and look thereon,' thus bringing new meaning to the term 'headlines'.

Ciphers also started to be used in classical times. A cipher changes a message on a letter-by-letter basis, from the so-called plaintext to the ciphertext, and back again. A code, on the other hand, converts whole words or phrases into other words or numbers. Julius Caesar used an early cipher system for his private communications. He took the plaintext and simply replaced each letter by the one that came a fixed distance away in the alphabet. For instance, *A* might be replaced by *B*, *B* by *C*, and so on. In the current English alphabet, this simple method of substitution gives rise to twenty-five different ciphers. The recipient needs only to be told which Caesar cipher is being used in order to convert back to the plaintext.

Needless to say, Caesar ciphers aren't very secure. In fact any type of monoalphabetic substitution, where each letter of the alphabet is replaced according to a key with another letter or symbol, is highly vulnerable. Sherlock Holmes shows how to go about unravelling such a cipher in 'The Adventure of the Dancing Men'. The wife of his client begins receiving messages in the form of rows of stick figures in various poses. Starting from the assumption that each different figure stands for a different letter, he quickly establishes which one stands for *E* – the commonest letter in the English language. Then he works his way through the next most common letters, *T*, *A*, *O*, and so on, trying different combinations until he has a few partial words deciphered. This allows him to fill in missing gaps with educated guesses. For instance, '*T_E*' is most likely to be '*THE*', which then gives him '*H*'. Holmes's task becomes easier as more of the dancing men messages are supplied to him.

Mary, Queen of Scots famously used ciphers to communicate with her co-conspirators in plotting the death of her cousin, the Protestant queen of England, Elizabeth I.

Mary's letters, encrypted by her secretary, Gilbert Curle, were smuggled out of Chartley Manor in Staffordshire, where she was being held prisoner, on the stoppers of casks of ale. Cryptography had advanced a long way since Caesar's time, and the ciphers in which Mary's communications were sent substituted not only symbols for letters but also for some words. They included 'nulls' as well – symbols that stood for nothing at all – intended to throw off anyone trying to interpret them as letters, and a *doubleth*, to indicate that the next character was to appear twice. But just as the art of ciphering had progressed by the second half of the sixteenth century, so had that of deciphering. Unbeknown to Mary, the courier of her covert communiqués was a double agent working for Elizabeth's principal secretary and head of intelligence, Sir Francis Walsingham. Everything she wrote was intercepted and handed over to Walsingham who, in turn, passed the messages on to Sir Thomas Phelippes, a Cambridge-educated linguist and master codebreaker. Mary's fate, and that of the whole plot against Elizabeth, was sealed when she sent a reply to Sir Anthony Babington, a leader of the conspiracy, which clearly implicated her. Phelippes deciphered the letter and then shrewdly amended it before retransmitting it, as if from Mary, to Babington, asking for the names of the others involved. All were summarily rounded up and executed. Mary incriminated herself by writing: 'Orders must be given that when their design has been carried out I can be...got out of here.' After a brief trial she was condemned to death by beheading.

Mary would have been better served, had she known about it, by using a far more elaborate cipher system invented in 1553 by Italian cryptologist Giovan Bellaso. It's based on polyalphabetic substitution, in which a number of different alphabets are used to encrypt the plaintext. The idea of deploying multiple alphabets had been hatched almost a century earlier, in 1466, by architect, founder of the science of projective geometry, and all-round intellect Leon Alberti. In Alberti's scheme the switch to a different cipher alphabet takes place every few words and is flagged by the appearance of a key letter in the ciphertext. Encoding and decoding are done using a metal disc consisting of two concentric wheels – a fixed outer one and a moveable inner one – with a common central pin. The outer ring is divided into cells, one for each upper-case letter of the plaintext plus the numbers 1 to 4 used to indicate phrases contained in a codebook. The inner ring is inscribed with lower-case letters, in jumbled order, for the ciphertext.

In the early 1500s, German abbot and polymath Johannes Trithemius did away with the need for Alberti's disc. In its place he used what he called the *tabula recta* – a square made up of the 26 letters of the alphabet followed by 25 rows of additional letters, each shifted once to the left, to create 26 different Caesar ciphers. The ciphertext generated by Trithemius's method may have looked like a random string of characters but it was still vulnerable to frequency analysis (of the sort used in Holmes's adventure), especially if the encryption strategy were known. It also suffered from the fatal weakness that the points in the ciphertext where a switch was made to a new alphabet had to be indicated in the body of the message by index letters. Giovan Bellaso's breakthrough was to take this information out of the message, where it could be discovered, and, instead, identify the switch to other alphabets, which took place every letter, by means of an agreed-upon countersign or keyword. Each letter of the keyword identified the row in the *tabula recta* that contained the alphabet to be used in deciphering the next character of the message. The longer the keyword, the fewer times the cipher alphabets needed to be reused.

Whereas Alberti and Trithemius employed a fixed pattern of substitutions, Bellaso's scheme meant that the pattern of substitutions could easily be changed, simply by selecting a new key. Keys were typically single words or short phrases, known in advance to both sender and receiver.

Bellaso didn't get proper credit for his invention. That went instead to French diplomat and cryptographer Blaise de Vigenère, a near contemporary of Bellaso's, due to a misattribution in the nineteenth century. The Vigenère cipher, as it became known, gained a reputation for being very strong. Charles Dodgson (aka Lewis Carroll) described it as being unbreakable in a children's article called 'The Alphabet Cipher' (1868). But it was far from that. Some cryptanalysts could occasionally crack it not long after it first appeared, in the sixteenth century. In 1863, a definitive method for breaking the Vigenère cipher, along with all other polyalphabetic substitution ciphers that use a keyword to switch between alphabets, was published by German infantry officer and cryptographer Friedrich Kasiski.

Despite having fatal weaknesses, the Vigenère cipher saw plenty of action. During the American Civil War, the Confederate States used it, together with a brass cipher disc, to transmit messages between its leaders. Unfortunately, the Union regularly intercepted and deciphered these, partly due to the Confederates' repeated use of the same three key phrases: 'Manchester Bluff', 'Complete Victory', and 'Come Retribution'.

In the other main type of cipher, known as a transposition cipher, letters are rearranged rather than substituted according to a predetermined rule or key. The instruction might be, for instance, to swap around every pair of letters in the plaintext, or write all words backwards. The more complex the prescribed rearrangement, the more difficult it is for a codebreaker to understand what's going on. Searching for anagrams is an effective way of cracking a transposed message but it can take a lot of time. In both World War I and the American Civil War, long before computers were available to automate anagram searches, sensitive messages were often sent in transposed form.

Just after the end of World War I, German engineer Arthur Scherbius invented a remarkable encryption device called Enigma. It was eventually produced in a variety of models, both commercial and military, before finding its most

An Enigma machine, used during the late 1930s and World War II, on display in the Museum of Science and Technology, Milan.

notorious use – to encode secret German communications during World War II. Resembling a typewriter, the Enigma contained a series of three rings or rotors, each of which bore the letters A to Z. Above the keyboard was a lamp board with a small light for each letter. When the operator pressed the key for a letter of the plaintext, the corresponding enciphered letter lit up on the lamp board. Pressing a key caused the rings to turn and kept the cipher changing continuously.

In the lead-up to the war, British codebreaking attempts to crack Enigma relied mainly on linguists and had limited success. Far more progress was made in Poland, where mathematicians were deployed on the problem and built machines of their own to simulate the workings of Enigma. These 'bombas', as they were called, because of the ticking noise they made while working, enabled operators to search rapidly through many different settings until they hit upon the one that had encoded the message.

Nazi scientists, however, made security improvements to Enigma. A plug board, installed on the front of the machine, enabled pairs of letters to be transposed. This system, combined with the existing one of the rings, boosted the number of possible settings to around 160 million trillion. Every day, German commanders, who all had identical Enigma machines, were issued with a new initial wheel configuration so that they could decipher each other's messages. With these extra levels of security in place the bombas were rendered ineffective. But the Poles had shared their knowledge with the British, who assembled an expert team of codebreakers, including Alan Turing and other talented mathematicians, at Bletchley Park, Buckinghamshire. Over time, Turing and his colleagues developed their own 'Bombe' capable of deciphering the more complex wartime Enigma codes.

The Bombe's task was made easier by other cryptanalysts at Bletchley who identified 'cribs' – partial translations of messages arrived at through a combination of monitoring, analysis, and smart guesswork. For example, coded messages picked up from German weather stations were deemed likely to contain an encryption of the term 'Wettervorhersage' (weather forecast) at a similar place in each message. Another clue came from the fact that the Enigma couldn't encrypt a letter as itself, so the encrypted message could be lined up in different ways with the crib until there were no alignments of the same letter.

Turing's Bombe was a formidable piece of engineering. Seven feet wide, six-and-a-half feet high, and weighing a ton, it contained twelve miles of wiring and 97,000 different parts. In effect, it was a bank of 36 Enigmas, replicated down to the individual wire. At the start of a run, each cloned Enigma was supplied with a pair of letters from a crib obtained that day (for example, when a *P* became a *C* in the guessed word) and its three rings rotated to check all possible 17,576 (26^3) positions until it found a match. When each of the Enigma clones hit upon what appeared to be the correct letter pairing at the same time, the Bombe stopped and produced its output. Analysts then used the Bombe's findings, together with other information, to figure out the key that the Germans were using on that day. Knowing this, they could set up an Enigma with this key and decrypt every message intercepted during that 24-hour period.

After the Bletchley prototype proved successful, a couple of hundred bombes were built and stationed in various places around Britain to reduce the chance of this critical asset being destroyed in air attacks. Running around the clock the bombes decoded up to 3,000 German messages a day – a total of around 2.5 million by the end of the war – giving the Allies invaluable intelligence on enemy movements and strategy.

Military and diplomatic uses always spring first to mind when it comes to codes and ciphers, whether in fact or fiction. What espionage novel of the Cold War era would be complete without reference to spies and their codebooks? Each number or code word in a secret message could be translated only by looking up its plaintext equivalent in a codebook – a kind of dictionary – which had to be kept from enemy eyes at all costs. In real life, different countries and organisations did indeed have their own unique codes and changed them regularly to avoid the content of sensitive communications being intercepted and decrypted through techniques such as word frequency analysis. But all of these 'private-key' systems were clumsy to use and vulnerable to the codebook falling into the wrong hands.

In 1973, a much more powerful technique of encipherment emerged, having been developed secretly at GCHQ, the UK's government surveillance centre (a fact not acknowledged until 1997). Known as public-key encryption, it's now universally used whenever we order something online, transfer funds electronically, or even make a Skype video call. It's based on the fact that it's a lot harder to work out the factors of a number (7 and 2, for example, are factors of 14) than it is to multiply them together. If the two factors are huge prime numbers, a computer can almost instantly times them together but, starting only with this product, be unable to figure out the factors in any reasonable length of time.

Public-key encryption uses a pair of keys: one public and one private. The public key is a gigantic number that's the product of two big primes, which together make up the private key. A mathematical formula allows the public key to be used to encrypt a message. Only the holder of the private key, however, can readily decrypt the message and extract the information it contains. As the speed of computers has increased so has their ability to do 'brute force' searches for the factors of the large numbers used in the public key. This has led to the adoption of new algorithms and of even larger public keys, which are almost impossible to break down into factors by working through all conceivable combinations. Hackers, however, can use other approaches to circumvent the encryption, such as applying the power of high-speed processors to try to crack user passwords.

Even with all the know-how and technological tools at their disposal modern cryptanalysts have been stumped by several famous codes and ciphers. One of these was devised by Ron Rivest, inventor of the algorithm that's most commonly used for public-key cryptography. In 1999, Rivest hid a mystery prize inside a metre-tall lead casket in the computer science lab at MIT. The casket has a time lock that won't open until 2033, on the lab's seventieth anniversary, unless someone manages to solve earlier the fiendishly hard challenge that Rivest has set. His problem starts by asking for the remainder when one number that's 7,200 trillion digits long is divided by another that's 616 digits long. This remainder, which also has 616 digits, is the key to unlocking the code, but only after it's been converted to binary – a string of 0s and 1s – and compared with the binary form of the original 616-digit number. Whether a breakthrough in computational speed or algorithmic design will allow someone to release the time lock, earlier than the thirty-five years Rivest has predicted, remains to be seen.

Another unsolved puzzle stands outside the CIA's headquarters in Langley, Virginia. Erected in 1990, it's a copper sculpture called *Kryptos* by American artist James Sanborn. The sculpture bears 1,735 encrypted letters in four sections, three of which have been decoded, yielding enigmatic messages that suggest some broader mystery connecting the whole piece. Tantalisingly, though, the final section has so far defied all attempts to crack it.

More sinister are the coded letters sent to local police and newspapers by a serial killer in San Francisco in the late 1960s. The killer, adopting the moniker 'Zodiac', murdered at least seven people and claimed to have hidden clues to the crimes in several encrypted messages. Three of the messages were disguised by substituting symbols for letters but with an added twist: multiple symbols were used for the commonest letters, such as 'e', 't', and 'a', making these hard to recognise by the usual method of frequency analysis. The three messages were eventually deciphered by looking for the words 'kill' and 'killing' and turned out to include a description of the perverse pleasure the murderer gained from the deaths. One of the last coded communications from Zodiac to local papers used a different encryption method. Although law-enforcers suspect it may hold vital clues to Zodiac's identity, its contents remain a mystery.

Sometimes, as with the Voynich manuscript, it's hard to tell if what looks like a code is actually solvable or merely a hoax. In others, it may be that the code is valid but the contents are misleading or fictional. In 1885, so the story goes, an adventurer named Thomas Beale buried two wagonloads of silver coin in Bedford County, near Roanoke, Virginia. He gave three encoded written messages to a friend supposedly telling where the treasure was hidden then disappeared into the west and was never heard from again. Several years later, the code in the second letter was cracked and turned out to be based on text from the Declaration of Independence. A number in the letter indicated which word in the document was to be used. The first letter of that word replaced the number. For example, if the first four words of the document were 'We hold these truths', the number 3 in the letter would represent the letter *t*. The second letter starts:

> I have deposited in the county of Bedford about four miles from Bufords in an excavation or vault six feet below the surface of the ground the following articles belonging jointly to the parties whose names are given in number three herewith. The first deposit consisted of ten hundred and fourteen pounds of gold and thirty-eight hundred and twelve pounds of silver deposited Nov eighteen nineteen...

According to one theory, both the remaining letters are encoded using either the same document in a different way, or another very public document. But, the fact is, we don't know and, in the end, the Beale ciphers could simply amount to an elaborate and entertaining wheeze.

The maths behind modern cryptography is all that keeps secure the bulk of today's financial transactions. It guards the privacy of our online communications. It rests on the principle that whereas some maths operations are easy to do in one direction, they're fantastically hard to do in reverse, so that, without a private key, it's almost impossible to unlock a code once generated. *Almost* impossible. Mathematicians know that they can never rest in the game of cryptographic cat and mouse. As computers become ever more powerful, and quantum computers come of age, ingenious new methods will be needed to keep the world's secrets safe.

CHAPTER 4

Fantasia Mathematica

I think there's something heavenly about numbers, anyway.

– Agatha Christie, *The Moving Finger*

MATHS IS ALREADY fantastic. You'd be hard pressed to dream up anything as outrageous as different sizes of infinity or shapes in 56 dimensions. Write a story in which a ball is cut apart and then reassembled to make two balls each as big as the original and it'd seem far-fetched – except that it's a genuine mathematical result. Even the characters and incidents you might contrive to bring a maths novel alive couldn't be more outlandish than some of those in real life. Maths often finds its way into fiction *because* it's so fantastic and makes for a good plot device. At other times an invented story serves to explain some aspect of maths or to speculate where maths might lead in the future.

Scientists who also wrote science fiction, such as Carl Sagan and Fred Hoyle, have put forward ideas in their stories that lie far beyond the bounds of established fact – ideas they may have been reluctant to air in professional circles. In the same way, authors have used tales of the imagination to explore possible new realms of mathematics. Isaac Asimov based his *Foundation* trilogy on the premise that a future field of maths, called psychohistory, could accurately foretell the actions of large groups of people.

In no branch of science has a major breakthrough first been widely and accurately disseminated in a work of fiction. But this has happened in the case of number theory. Computer scientist Donald Knuth's novella *Surreal Numbers: How Two Ex-Students Turned On to Mathematics and Found Total Happiness* was the first published work to describe an important new system of numbers (discovered by John Horton Conway). The ploy of explaining scientific ideas in a fictional context, though, is an old one, and crops up regularly in what may be thought of as a subcategory of science fiction: mathematical fiction.

As long ago as 414 BCE, maths appears in a comedy called *The Birds* by the Greek playwright Aristophanes. At one point an actor, portraying the geometer Meton of Athens, comes on stage carrying some surveying instruments and explains:

> With the straight ruler I set to work to inscribe a square within this circle; in its centre will be the marketplace, into which all the straight streets will lead, converging to this centre like a star, which, although only orbicular, sends forth its rays in a straight line from all its sides.

Fast-forward to 1666 and maths pops up again in one of the earliest works of science fiction and possibly *the* earliest written by a woman. In a passage of *The Blazing World* by Margaret Cavendish, Duchess of Newcastle, the heroine of the tale is being introduced to some of the inhabitants of another planet. Two of the intelligent species there, the lice-men and the spider-men, turn out to be excellent mathematicians and keen to explain their achievements. Some sixty years later, in *Gulliver's Travels*, Jonathan Swift describes the encounter between his protagonist and the Laputians who are interested in only two things: mathematics and music. Such is the depth of their obsessions that all their food is served in the shape of either a mathematical figure or a musical instrument.

The Victorian era saw a rapid growth in the appearance of maths in fictional settings. In 1852, Charles Kingsley, author of the children's classic *The Water-Babies*, wrote an entire novel about the life of Hypatia, the first female mathematician whose work is reasonably well recorded. As science fiction became established as a distinct genre, at a time of rising industrialisation and popular interest in scientific and engineering breakthroughs, so maths began to seep through more and more into the work of authors such as Edgar Allan Poe, Edward Page Mitchell, and Jules Verne. Among the most mathematical writers was Charles Dodgson, better known by his pseudonym Lewis Carroll, who had a first-class degree in maths from Oxford and later became a lecturer in the subject at Christ Church.

The second half of the nineteenth century was an exceptionally fertile time in mathematics. Radically new ideas were fast evolving in areas like non-Euclidean geometry, abstract algebra, and complex numbers. Not surprisingly, given that he lived at a time of such intellectual flux, Carroll's books are filled with colourful mathematical allusions and challenges to conventional ways of thinking. Although generally considered a mathematical conservative, more at home with Euclid's *Elements* than the seismic shifts in the subject taking place all around him, Carroll still comes across as a bohemian in spirit.

The Mad Hatter, illustration by Sir John Tenniel (1865).

'What's the difference between a raven and a writing desk?' the Mad Hatter asks Alice. When challenged, the Hatter admits he has no clue. 'I think you might do something better with the time than wasting it in asking riddles that have no answers,' replies the exasperated Alice. Puzzle enthusiast Sam Loyd came up with his own solution: 'Poe wrote on both.' As for Carroll himself, so often was he quizzed about the real answer that he finally invented one: 'Because it can produce a few notes, tho they are very flat; and it is nevar put with the wrong end first!' Unfortunately, before his explanation appeared in print, a proofreader, unaware that Carroll had intentionally written 'nevar' – 'raven' in reverse – corrected the spelling so that some of the original wit was lost.

Carroll interspersed his fantasy writings with factual books on maths and so began a trend that's continued to this day. Among his strictly mathematical works are *A Syllabus of Plane Algebraic Geometry* (1860) and *Symbolic Logic* (published posthumously). Often he combined fact and fiction in an informal and playful style, as in *The Game of Logic* (1887), in which he explains logical propositions and inferences by way of a board game. In *Euclid and His Modern Rivals, A Tangled Tale* (1879), he argued that the two-thousand-year-old *Elements* was still the best text for teaching geometry. This wasn't a controversial claim in mid-Victorian times given that Euclid's classic treatise was second only to the Bible in the number of editions published since its first European printing in 1482. But Carroll's defence was unusual in that it took the form of a play featuring Euclid's ghost and the fictional characters 'Minos' and 'Dr Niemand'. Carroll's last novel, published in two parts, *Sylvie and Bruno* (1889) and *Sylvie and Bruno Concluded* (1893), was a mixture of fairy tale and social commentary. In the second book, the characters discuss, over tea, how to make a projective plane by gluing the edge of a disc to that of a Möbius band – a convoluted digression that probably contributed to the book's lukewarm reception.

In the same era, one of the great early popularisations of mathematics was penned by English schoolteacher and theologian Edwin Abbott. In reality, Abbott's main goal in writing *Flatland: A Romance in Many Dimensions* (1884) was to comment on the gross inequalities in Victorian society, especially with regard to women, its hierarchy being reflected in the geometry of Flatland's inhabitants. The females were mere lines and the lower-class males acute-angled triangles. Males of greater social standing took the form of equilateral triangles and, of greater importance still, polygons with more and more sides until they were almost indistinguishable from the most highly regarded shape of all – the circle. The enduring popularity of the book, however, is due to its user-friendly introduction to the maths of different numbers of dimensions rather than to its now-dated satire and cultural allegory.

The spirit and theme of Abbott's original *Flatland* have been explored further by other authors, right up to the present day. A. K. Dewdney's *The Planiverse* (1984) takes the ideas of *Flatland* to a new level – but not dimensionality – by imagining what science and engineering would be like in a 2D world. Dewdney, a Canadian mathematician and computer scientist, begins his tale with a class of computer science students attempting to simulate what life in two dimensions would be like, complete with workable physical laws and a functioning ecosystem. They're startled to receive a message, via the computer, from an inhabitant of this 'planiverse', who calls himself YNDRD, pronounced by the students as 'Yendwed' ('Dewdney' spelled backwards). They go on to learn all about the goings-on in Yendwed's world, from the way digestive tracts work in creatures that, to us, are mere plan-views, to the types of molecules that are possible when one of our normal dimensions is missing.

Ian Stewart, mathematician and prolific author of popular maths books, wrote *Flatterland* (2001), which he explicitly intended as a sequel to Abbott's classic, albeit published more than a century later. His imagined narrator is the grand-daughter of 'A. Square', a member of the educated caste of gentlemen and professionals through whose voice Abbott told the original tale. Less dense, far shorter, and, indeed, barely mathematical at all, is Norton Juster's bon mot, *The Dot and the Line: A Romance in Lower Mathematics* (1963). Written for children, it tells the story of a straight line who falls hopelessly in love with a beautiful dot. The dot, how-ever, has eyes only for less sensible fellows in the form of squiggles until our hero learns all about angles and manages to transform himself into far more interesting and alluring shapes. A couple of years after it was published, *The Dot and the Line* was adapted into a ten-minute short film by famed animator Chuck Jones for Metro-Goldwyn-Mayer. Narrated by the English actor Robert Morley, almost verbatim from the book, it won the 1965 Academy Award for Best Animated Short Film and was one of only two animations released by MGM that weren't Tom and Jerry cartoons.

At the opposite end of the difficulty scale from *The Dot and the Line* was Charles Howard Hinton's *An Episode of Flatland* (1907). Hinton was a curious and colourful character – briefly a bigamist, inventor of an ingenious but dangerously unreliable baseball gun, and lifelong obsessive about higher dimensions. He wrote several books and essays on the fourth dimension and even devised a collection of coloured wooden blocks, sold commercially and by which, he claimed, he taught himself to see in 4D. In *An Episode of Flatland*, however, he descends a dimension to the plane world of Astria. Borrowing heavily from Abbott in overall concept, he does a better job of explaining the physics of his 2D world, though the plot is tedious and the characters unremittingly flat.

By the late nineteenth century, mathematics, in one form or another, was making regular appearances in fiction of all types but especially that dealing with technical issues. Conan Doyle brought it into his mysteries, most notably when his master detective, Sherlock Holmes, needed to draw on a knowledge of geometry or cryptography. Later authors, however, have noted a few maths and science bloopers in the Holmes tales. In 'The Final Problem', Holmes says of his nemesis, Professor James Moriarty: 'At the age of twenty-one he wrote a treatise upon the Binomial Theorem, which has had European vogue.' As Isaac Asimov pointed out in a chapter of his book *The Roving Mind* (1983), Moriarty would have been twenty-one years old in about 1865. This was forty years after Norwegian mathematician Niels Henrik Abel had fully worked out every aspect of the binomial theorem, leaving Moriarty with nothing new to find about the subject.

H. G. Wells, who, along with Jules Verne, was one of the early heavyweights of science fiction, delved into mathematical descriptions in several of his novels. Early in *The Time Machine*, the time traveller announces to his colleagues:

> You must follow me carefully. I shall have to controvert one or two ideas that are almost universally accepted. The geometry, for instance, they taught you in school is founded on a misconception.

He then goes on to explain how a four-dimensional geometry would work and adds that 'Professor Simon Newcomb was expounding this to the New York Mathematical Society only a month or so ago.' In fact, American astronomer Newcomb had indeed lectured on the fourth dimension to the New York Mathematical Society in December 1893. It was this very lecture that inspired Wells to imagine, in his 1895 novel, a time-travelling device, moving in 'another dimension at right angles to the other three'.

Wells returned to the theme of higher dimensional excursions in a short tale called 'The Plattner Story', published a year after *The Time Machine*. Gottfried Plattner, the central character, is a young teacher in a prep school in Sussex. One day, bored while supervising a group of boys doing their evening homework, he decides to test a mysterious green powder that one of his pupils has brought in for analysis. Plattner knows next to nothing about chemistry – he teaches mainly modern languages – and so tries various tests at random. After the addition of nitric, hydrochloric, and sulphuric acids fails to elicit a response from the powder, he puts a match to a sizeable pile of it. The next moment there is a huge explosion that blows out the window of the classroom, sends the students scurrying under their desks in fear for their lives, and, it seems, causes Mr Plattner to vanish into thin air.

For more than a week the mystery of Plattner's disappearance endures. Then, just as suddenly as he'd winked out of existence, he comes back, crashing with a thud into the headmaster of the school, who is out in his garden doing some weeding. Over the next few days, a bizarre fact emerges – everything about Plattner has been swapped right to left. He's become the exact mirror image of his old self: left-handed instead of right and with all his internal organs – liver, lungs, and so forth – switched around. Just as we could pick up a two-dimensional right-handed glove, flip it over in a third dimension, and put it back down to become a left-handed glove, so 'the curious inversion of Plattner's right and left sides is proof that he has moved out of our space into what is called the Fourth Dimension.'

The pulp era of science fiction magazines, which began in the US in the 1920s, and the 'Golden Age' of SF, which followed and overlapped with the years of World War II, produced a torrent of fantastic tales about the future. These imaginings ranged from the truly awful to the inspiringly creative, and were marked by a shift away from the often long-winded romanticism of Victorian and Edwardian literature to 'hard science fiction' with a focus on theories, technology, and scientific veracity. A smattering of maths, at least, found its way into many of these science-based tales, in the guise of adventures in a fourth or fifth dimension, calculations needed for faster-than-light travel, or musings on infinity. Some of the new hard SF had mathematics as its central theme.

In Nathan Schachner's *The Living Equation* (1934), a mathematician builds a machine that's designed to transform purely abstract objects, such as vectors and tensors, into physical reality. But before he has a chance to test the invention, it's inadvertently tripped into action by an intruder at his home. The consequences are disastrous: whole buildings move or disappear, their inhabitants suddenly finding themselves plunged into an alternative dimension; land masses vanish; oceans are swallowed up; and time slows down or speeds up in different parts of the world. Behind the story is a serious philosophical notion, namely, that maths represents the ultimate reality in which we're embedded, while the physical universe is merely an illusion that dances to whatever mathematical tune is being played. Greg Egan explored similar territory in a 1995 short story. 'Luminous' tells the tale of two graduate maths students who make the astounding discovery that what had been thought of as ultimate truths in number theory are really far more local and temporary than expected – changeable over time and in competition with other 'truths' that apply elsewhere in the infinite reaches of the numerical cosmos.

Mathematics and philosophy go hand in hand, too, in some of the short stories of Argentinian writer Jorge Luis Borges. 'The Library of Babel', originally published in Spanish in 1941, supposes the universe to consist of a vast library containing all possible books of a standard size (410 pages), format, and character set, with twenty-two letters, the full stop, the comma, and the space. Every combination of these characters, subject to the other constraints, appears somewhere in the library. Inevitably, only a tiny fraction of the books contain even short passages that make sense. On the other hand, everything that has been written or ever could be written, including everything fictional (even if it purports to be true) and factual, crops up somewhere in the mind-bogglingly large – and utterly useless – collection.

Huge numbers of combinations of symbols are also central to the plot of 'The Nine Billion Names of God' (1953) by Arthur C. Clarke. The monks of a Tibetan monastery hire a computer and two programmers to speed up their task of encoding, according to their belief, all the possible names of God, each of which consists of no more than nine characters. Doing it by hand they reckon would take about 15,000 years. Electronic automation ought to drastically reduce this and the monks are keen to get the job done because then, they say, God will bring the world to an end and we can get on with enjoying whatever comes next. Sure enough, the computer zips through its task in just a few months and the programmers head for home, leaving the monks to paste the final few names into their holy books. As the Westerners make their way along a mountain path towards the airfield and their journey back to civilisation, confident that their task though complete has been mercifully futile, they pause to look up only to see that 'overhead, without any fuss, the stars were going out.'

Clarke, Isaac Asimov, and George Gamow (like Asimov a Russian émigré who became an American citizen) were prominent among qualified scientists who, beginning in the 1940s, wrote essays and books on both science fiction and science fact. Sometimes they fused the two, as in the case of Gamow's four books featuring Mr Tompkins, a bank clerk engaged to the daughter of a famous physicist, whose dreams transport him into realms where physical constants are altered so that he can better understand relativity, the world of the atom, and modern cosmology. In the same era, especially in the post-WWII years, writers with a strong mathematical background started publishing popular material that sometimes crossed over between fact and fiction.

Pre-eminent among the modern wave of maths popularisers was Martin Gardner, an American writer with a bachelor's degree in philosophy from the University of Chicago and a keen interest in maths puzzles, magic tricks, and the debunking of pseudoscience. His articles on recreational maths, which appeared monthly in the 'Mathematical Games' column of *Scientific American* for about a quarter of a century, beginning in December 1956 with a piece called 'Flexagons', encouraged a generation of young people to enter the field and enthralled many others, from professional mathematicians to interested bystanders. Such was Gardner's stature that world-class specialists, such as John Conway, Richard Guy, Donald Knuth, and Roger Penrose, corresponded with him regularly and saw their work explained in an easy, enticing style to millions of readers, who might otherwise have never heard of Conway's Game of Life (and other cellular automata), polyominoes, the Soma cube, Penrose tilings, or the superellipse. As Gamow had done earlier, Gardner adopted a fictional character, Dr Matrix, to add spice, humour, and a little satire to some of his mathematical explanations. Dr Irving Joshua Matrix was first introduced in Gardner's *Scientific American* column in January 1960 and went on to make eighteen appearances in all before taking his final bow in the September 1980 article titled 'Dr Matrix, like Mr Holmes, comes to an untimely and mysterious end'.

Such was the amount of popular maths-related material, both factual and fictional, that had built up, even before the likes of Gardner got into their stride, that American editor,

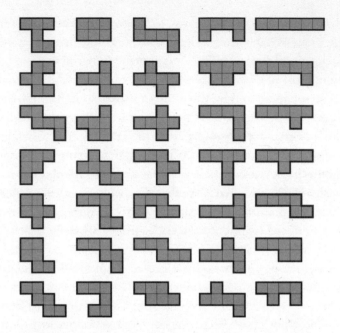

Solomon Golomb's polyominoes were among the many recreational mathematics topics featured by Martin Gardner in his *Scientific American* column. Shown here are all 35 hexominoes.

author, and early media personality Clifton Fadiman was encouraged to pull together an anthology of such writings in his *Fantasia Mathematica* (1958). Forty years later, William Frucht, an editor with whom David has worked, composed a similar collection of mathematical articles, short stories, poems, and other whimsy. He called it *Imaginary Numbers* and wrote it after failing, on various occasions, to persuade Gardner to take on the task.

Popular maths literature, much of it injected with elements of humour and fantasy, began to mushroom in the late 1970s and early 1980s. American professor of cognitive science Douglas Hofstadter won a Pulitzer Prize for his *Gödel, Escher, Bach: An Eternal Golden Braid* (1979), a book that explores common themes in the works of logician Kurt Gödel, artist Maurits Escher, and composer Johann Sebastian Bach. In his July 1979 'Mathematical Games' column, Martin Gardner wrote of it: 'Every few decades, an unknown author brings out a book of such depth, clarity, range, wit, beauty and originality that it is recognized at once as a major literary event.'

Hofstadter employs an unlikely duo who were first called upon to explain some tricky concepts to the public two and a half thousand years earlier. Achilles and the Tortoise debuted as a double act in the writings of philosopher Zeno of Elea, famed for his paradoxes that bore upon the nature of infinity. The best known of these is illustrated by a handicap race between the swiftest runner of ancient Greece and the slowest moving of reptiles, which, the Tortoise argues (compellingly but wrongly!) he's bound to win. Lewis Carroll recalled the pairing in 'What the Tortoise Said to Achilles', an article he contributed to the philosophical journal *Mind* in 1895. Again, the Tortoise proves that in mental agility, at least, he's more than a match for his heroic opponent by outwitting him in a debate on logic that leaves Achilles floundering in an infinite regress. Hofstadter reprises the combo, having them engage in numerous lengthy dialogues throughout *Gödel, Escher, Bach*. The two are joined by other characters including the Crab, the Anteater, and the Sloth, in chatting about and illustrating self-reference, formal rules in logic, how knowledge can be represented and stored, and the emergence of consciousness from unconscious matter. *GEB* may rank among the most technically dense books on maths and logic ever written for a general audience. Most people will need to reread it and study it almost like a textbook to grasp properly what it's all about. But it succeeds – even in explaining such esoterica as Gödel's incompleteness theorems – through its clever use of playful ideas and fanciful characters who speak plain, everyday language.

The success and popularity of authors like Gardner and Hofstadter encouraged other mathematicians to try their hand at writing about their subject in a whimsical way. It's an approach that works because it counterbalances the common perception, often gained in school, that maths is dry and difficult. Rudy Rucker taught mathematics for several years in the 1970s at the State University of New York at Geneseo, then lectured in Heidelberg and Virginia, before spending the rest of his academic career, until retirement, as a professor of computer science at San José State University. Rucker was an early contributor to the cyberpunk movement and wrote prolifically about maths both in fiction and non-fiction for the layperson. In *Pi in the Sky*, a couple skin-diving on their honeymoon come across an artefact from Barnard's Star, lost on the seabed during the crash of an alien spacecraft. Cone-shaped and marked with countless black rings, it gives off an intricate pattern of hisses. The device turns out to be a kind of portable hard drive that holds all the knowledge of the Barnardians in a fantastic form – coded in the decimal expansion of pi. The same idea found its way into the plot of another, and much better-known story, a couple of years later. 'My notion of pi as a kind of universal library', remarked Rucker, 'reappears in Carl Sagan's novel *Contact* in 1985'.

Rucker also brought fanciful elements into his non-fiction, including *The Fourth Dimension* and *Infinity and the Mind*. In a similar vein, Clifford Pickover, Ian Stewart, Alex Bellos, and others of the current crop of popular maths writers often use playful examples, invented characters, and fantastic settings to make accessible topics that might otherwise seem remote and hard to fathom.

In recent years, the lives of several famous mathematicians have become better known through popular biographies that, in some cases, have been adapted as films. *A Beautiful Mind* (1998), by Sylvia Nasar, recounts the troubled life of Nobel Prize-winning economist and mathematician John Forbes Nash who battled schizophrenia. It inspired the 2001 Oscar-winning movie of the same name, directed by Ron Howard and starring Russell Crowe as the central character. Ramanujan, too, has made it onto the big screen in the 2015 drama *The Man Who Knew Infinity*, based on the 1991 book with the same title by Robert Kanigel.

Rain Man (1988), likewise based on a true story, co-stars Dustin Hoffman as an autistic savant with a photographic memory and a genius for mental arithmetic. *Good Will Hunting* (1997), written by Matt Damon and Ben Affleck and starring Robin Williams, is about a young man who's led a troubled life but has an amazing talent for maths. His abilities are discovered when he falls foul of the law, and soon after he has to decide whether to focus on making the most of his mathematical genius or carrying on with his old lifestyle. In Darren Aronofsky's disturbing independent film *Pi* (1998), the main character is a mathematician obsessed with searching for patterns in pi's infinite decimal places. He believes these patterns could be used to predict chaotic behaviours, including that of the stock market. Throughout the story he's pursued by ruthless financial players and by religious zealots hoping to find a mathematical way to communicate with God.

Maths has also found its way onto the stage. The musical *Fermat's Last Tango* (2000), performed in New York by the York Theatre Company, is a fictionalised account of the struggle of Andrew Wiles to prove (successfully in the end) Fermat's last theorem. It followed the Pulitzer Prize-winning play *Proof* by David Auburn about the death of a brilliant mathematician and the repercussions for his daughters and student.

Sometimes mathematics is made more palatable and human by telling the story of those who pioneered it, or by presenting it in a fictional setting. But, in truth, maths is by nature a profoundly human endeavour, no less fantastic and fun in reality than the tales that have been woven around it.

CHAPTER 5

In Beauty Lies Truth?

> All mathematicians share…a sense of amazement over
> the infinite depth and mysterious beauty and usefulness
> of mathematics.
>
> – Martin Gardner

EVERYONE CAN APPRECIATE beauty in art, music, poetry,
and nature. We're all capable of experiencing beautiful things,
even though our tastes and opinions may differ. *But beauty
in maths?* To many people, even the sight of an equation is
enough to stir negative thoughts and feelings – of a subject
that they never enjoyed or could properly understand. They
might find it hard to empathise with Bertrand Russell who
said: 'Mathematics, rightly viewed, possesses not only truth,
but supreme beauty – a beauty cold and austere, like that of
sculpture.'

Russell's sentiment has been echoed time and again by
mathematicians over the years. Georg Cantor, who opened
the door to our understanding of infinity, remarked: 'The
mathematician does not study pure mathematics because it
is useful; he studies it because he delights in it and he delights
in it because it is beautiful.' Polish mathematician Stefan

Banach, founder of the field of functional analysis, went a step further: 'Mathematics is the most beautiful and most powerful creation of the human spirit.'

A team of neuroscientists at University College London (UCL) looked into the brains of fifteen professional mathematicians to see what was happening when they were thinking about different equations. Prior to the tests, involving an MRI scanner, the mathematicians were asked to rank sixty well-known formulae on a scale ranging from ugliest (-5) to most beautiful (+5). The results of the scans were fascinating. They showed that the same part of the brain – a region known as the medial orbitofrontal cortex (OFC) – was active in mathematicians when they experienced what they claimed was beautiful maths as in other people who experienced what they described as beautiful art or music.

Earlier research had found that the OFC lights up in brain scans when someone is thinking about what they consider to be an attractive face, especially when that face is smiling. In the case of heterosexual women and homosexual men, activity was greatest for what was deemed an attractive male face, whereas the reverse was true for the opposite sexual orientation. Other studies have concluded that the OFC gets busy when its owner looks at what they characterise as a beautiful painting or listens to music – whether Vivaldi or Van Halen – that sends shivers up the listener's spine. It's this same part of the brain, lying just behind and above the eye sockets, which springs into action when the subject has mathematical beauty on their mind.

In any aspect of thinking, different parts of the brain are involved. But it's interesting that this same region, the medial orbitofrontal cortex, is such a focus of activity when responding both to the attractiveness of an equation and of a human face. Maths gives the impression to many people of being purely intellectual and dry, whereas beauty in art and nature is considered passionate and warm. Yet in objective measurements of the brain, both experiences fire up the same neurological structure. In mathematicians, the experience of beauty in equations triggers the same complex of neurons that in the population as a whole is activated by powerful emotional, and even sexually oriented, reactions to certain sights and sounds. Evidently, in terms of allure, mathematicians respond to what they consider to be attractive equations in the same way as people in general do to an attractive face and smile.

As far as the brain is concerned, it seems that the experience of beauty doesn't have to come from a sensory stimulus. It can, in some cases, be purely abstract. All that matters is the intensity of the feeling that's evoked. But this raises the question of how an equation or some other mathematical object or concept could ever be considered so beautiful as to raise such passions. Somehow it must involve not just the piece of mathematics itself – least of all its appearance as a bunch of squiggles on a piece of paper. Rather, the emotional response must arise from thought processes that grasp the deep meaning of the formula (or whatever it is) and its connections to other parts of maths.

In the UCL study, the formula consistently ranked as the most beautiful was Euler's identity – a simple-looking relationship that's long held a fascination for mathematicians. It links five of the most important constants in maths with three basic operations in the following way: $e^{i\pi} + 1 = 0$. A 1990 poll by the journal *The Mathematical Intelligencer*, asking readers to vote for what they considered the most beautiful formula, also saw Euler's identity come out on top. Theoretical physicist Richard Feynman called it 'the most remarkable formula in mathematics'. Paul Nahin, professor emeritus at the University of New Hampshire, described it as being 'of exquisite beauty'. Stanford professor Keith Devlin wrote:

> [L]ike a Shakespearean sonnet that captures the very essence of love, or a painting that brings out the beauty of the human form that is far more than just skin deep, Euler's equation reaches down into the very depths of existence.

Euler's identity wins the mathematical beauty contest, it seems, by combining profundity with extreme simplicity. It also has one other vital quality necessary for it to be attractive: it happens to be true. A unique strength of maths is its ability to prove (or disprove) things absolutely. Mathematicians are able to know, beyond any shadow of a doubt, when they've found truth in their subject. They can rigorously prove it – and the proof thereafter remains valid and unchanging for all time.

Is beauty a guide to truth in maths? What we consider to be beautiful or attractive in a human face is only skin deep and certainly no indicator of what the person might be like inside. It may be that in some cases superficial beauty in maths is misleading and that some error will eventually be found in a result that, at first sight, seemed so appealing that it must be right. Mathematicians are cautious about jumping to conclusions and are never happy until they have a final, incontrovertible proof. Yet they often comment on the beauty they find in the structure and symmetry of their equations. It's a commonly held belief that while beauty is no guarantor of truth it's at least a first sign of it. In *A Mathematician's Apology*, G. H. Hardy wrote:

The mathematician's patterns, like the painter's or the poet's, must be beautiful; the ideas, like the colours or the words must fit together in a harmonious way. Beauty is the first test: there is no permanent place in this world for ugly mathematics.

Even in maths, beauty is subjective – a human judgement call. In the UCL study mentioned earlier, some formulae, including Euler's identity, the Pythagorean identity, and the Cauchy–Riemann equations, were considered very beautiful. At the other end of the scale, Ramanujan's infinite series and Riemann's functional equations were ranked among the ugliest, perhaps because they're quite complicated and difficult to interpret. But beauty and ugliness are relative and to some theorists nothing in mathematics is unattractive. Bertrand Russell wrote:

> Mathematics possesses not only truth but supreme beauty – a beauty cold and austere, like that of sculpture, without appeal to any part of our weaker nature, without the gorgeous trappings of painting or music, yet sublimely pure, and capable of a stern perfection such as only the greatest art can show.

To some, the belief in the universal beauty of maths extends to equations that describe the real world. American architect Richard Buckminster Fuller saw beauty as an acid test of his designs:

> When I'm working on a problem, I never think about beauty. I think only how to solve the problem. But when I have finished, if the solution is not beautiful, I know it is wrong.

The most outstanding and obsessive supporter of this notion of the importance of beauty in the equations of nature was Englishman Paul Dirac, one of the greatest theoretical physicists of the twentieth century. Dirac had a bizarre upbringing. His father, Charles, was a Swiss immigrant, a teacher of French, and a strict disciplinarian. His mother was from Cornwall and worked as a librarian in the family's hometown of Bristol. He had a younger sister, Betty, and an older brother, Felix. Some idea of his extraordinary childhood can be gathered from the fact that, at mealtimes, Paul was forced to eat with his father in the dining room and speak only in French, while the rest of the family spoke English in the kitchen. If Paul made even the slightest mistake in his use of language, he'd be punished by having his next wish denied, even if this meant, as it often did, not being able to go the bathroom.

His early education was first-rate. Even at primary school he began to learn technical drawing – a subject he

Paul Dirac in 1933.

studied for nine straight years. Later, at high school, he had a superb teacher in projective geometry, so that Dirac, through these two areas of expertise, acquired a powerful visual sense for solving problems. Later, he acknowledged that geometric rather than algebraic thinking dominated his work in physics:

> My research work was based in pictures. I needed to visualise things and projective geometry was often most useful, e.g. in figuring out how a particular quantity transforms under Lorentz transf[ormation]. When I came to publish the results I suppressed the projective geometry as the results could be expressed more concisely in analytic form.

In part no doubt due to the way he was raised, Dirac was an oddball. He was deeply private, taciturn, and showed virtually no empathy towards others. He spoke little and was both anti-philosophical and anti-religious. Throughout his life, he berated anyone who tried to mix science or maths with artistic sensibility. In the late 1920s, at the University of Göttingen, he met American physicist Robert Oppenheimer (later head of the Manhattan Project), whom, he'd heard, wrote poetry, and said to him:

> I do not see how a man can work at the frontiers of physics and write poetry at the same time. They are in opposition. In science you want to say something nobody knew before, in words which everyone can understand. In poetry you are bound to say something that everybody knows already in words that nobody can understand.

Having started out in engineering, at which he was brilliant but lacked any practical ability, Dirac was encouraged to take a degree in maths. He then earned a scholarship to Cambridge to study for his PhD. However, instead of pursuing research in relativity theory, which he'd have preferred, he was given a supervisor, Ray Fowler, who was a specialist in the equally new but, to Dirac, much less attractive, field of quantum theory. Despite having aesthetic misgivings about the area in which he now found himself immersed, Dirac began to reveal the true extent of his genius. While still a doctoral student, he effectively became a co-founder of quantum mechanics by marrying two seemingly incompatible treatments of the subject – Werner Heisenberg's matrix mechanics and Erwin Schrodinger's wave mechanics. They were, he proved, exactly equivalent – the one could be mathematically switched into the other.

A year after receiving his PhD, Dirac was instrumental in another remarkable reconciliation: that of quantum mechanics and the special theory of relativity. By merging these two great pillars of modern physics, he provided the first relativistic description of the electron. It was summed up in a single formula, which soon became known as the Dirac equation. It's the only equation – not as if one would expect there to many others! – to be carved in stone in Westminster Abbey. Set into the floor near Newton's tomb, but far from where Dirac lies buried next to his wife in Tallahassee, Florida, it reads, in deceptively compact form: $i\gamma.\delta\psi = m\psi$. American physicist Frank Wilczek has described it as being 'achingly beautiful'. Dirac felt the same about it, to the extent that he trusted an extraordinary prediction that it made, which no one else at the time believed. He realised that his equation not only explained the spin and magnetism of the electron, but it also indicated there should be an *antielectron* – a particle with a charge equal and opposite to that of the electron and that was effectively the electron's mirror image. The idea of antiparticles wasn't taken seriously at the time but just a few years later, in 1932, the antielectron – or positron – was found by Carl Anderson in cosmic ray experiments at the California Institute of Technology. Dirac's confidence in his equation, based largely on its aesthetic appeal, had been rewarded with experimental verification that antimatter exists.

Dirac's equation was the first step in developing quantum field theory (QFT) – the basis for the Standard Model of particle physics today and the theory that predicted the existence of the Higgs boson. To the end of his life, Dirac hated QFT or, more precisely, he hated the various mathematical schemes that had to be bolted on to it to make it conform with the real world. One of these modifications, known as renormalisation, was introduced around 1950 by Julian Schwinger, Richard Feynman, Freeman Dyson, and Sin-Itiro Tomonaga. It successfully prevented the equations of QFT from blowing up to infinity in various situations that arise when particles interact. Dirac regarded renormalisation as 'ugly and incomplete' and insisted that something must eventually come along to replace it.

In none of his many scientific papers did Dirac mention his passion for mathematical beauty. Only in later life, mellowed perhaps by years, marriage, and being the father of two children, did he begin to express himself more freely in the rare talks and interviews that he gave. Like Einstein, when he used the term 'God' it wasn't in any traditional sense but as a synonym for nature as a whole, or some overall guiding principle behind the universe. Yet he was moved to say: 'God used beautiful mathematics in creating the world.'

Most controversially, he also said:

> [I]t is more important to have beauty in one's equations than to have them fit experiment... It seems that if one is working from the point of view of getting beauty in one's equations, and if one has really a sound insight, one is on a sure line of progress.

His advice was not to give up on a theory just because it doesn't agree with observations or measurements:

> [T]he discrepancy may well be due to minor features that are not properly taken into account and that will get cleared up with further development of the theory.

The Dirac memorial plaque, bearing the Dirac equation, in Westminster Abbey.

Most practising scientists would have a real problem with Dirac's assertion that an equation's beauty should take precedence over whether the equation actually fits measured facts. It's true that mathematical physicists gain aesthetic and emotional pleasure from the continued success, say, of Einstein's theory of general relativity – widely regarded as one of the most attractive descriptions of nature. Many hope, too, that some form of string theory will eventually be able to provide an explanation for the behaviour of particles and forces on the smallest scale, partly because it's seen as being so mathematically appealing. Others sound a warning about getting so attracted by the siren call of mathematical beauty that we lose sight of the need to have our physical theories agree with what instruments tell us. Unlike pure mathematics, science isn't a quest for truth but instead for better and better agreement between observation and theory. English mathematician Michael Atiyah put it this way:

> Truth and beauty are closely related but not the same. You're never sure that you have the truth. All you're doing is striving towards better and better truths and the light that guides you is beauty.

A dynamic interplay between emotion and reason, passion and proof, runs through both mathematics and science. This is inevitably so, because mathematicians and scientists are human. As Karl Weierstrass, regarded as the father of modern analysis, said: 'A mathematician who is not also something of a poet will never be a complete mathematician.' A century later, Einstein echoed the same sentiment: 'Pure mathematics is, in its way, the poetry of logical ideas.'

This feeling that there's more to maths than mere logical consistency shows itself in different aspects of the subject. There's what might be described as beauty in method, when a proof is regarded as being particularly elegant. The elegance might stem from the proof being unusually concise or requiring few new assumptions over previous results. It might come from the fact that the proof reaches its conclusion in an ingenious or satisfying way or that it generalises easily so that it ends up, almost serendipitously, solving similar, related problems.

A theory is often regarded as especially beautiful if it combines stylish formulation with depth. Deep results may be ones that form bridges between what had seemed to be disconnected areas of maths or that show mathematical structures in a new and more penetrating light. An example of the latter is a theorem published by Carl Gauss in 1828 in his *Disquisitiones generales circa superficies curvas* ('General investigation of curved surfaces'). Gauss defined a quantity that measures the overall, or total, curvature of a surface. He was then astonished and delighted to find that this curvature – the Gaussian curvature – is an *intrinsic* property of the surface. A way to think of this is that if there were intelligent inhabitants of the surface, they could figure out the curvature solely from measurements, such as distances and areas, made within the surface. They wouldn't need to know anything about the higher dimensional space in which it was embedded. The normally reserved Gauss was so startled by this result that he called it his Theorema Egregium ('Remarkable Theorem').

A much more recent theorem, known as the modularity theorem, is considered by many to be beautiful because of how it links two branches of maths that at first appear very different: elliptic curves and modular forms. Elliptic curves are smooth, plane curves generated by equations of the form $y^2 = x^3 + ax + b$. Modular forms are a special type of function that can explain many surprising and pleasing identities in number theory, such as those involving sums of squares. They're also used in two-dimensional modelling, and can be helpful, for instance, in describing molecular structures. The modularity theorem was originally known as the Taniyama–Shimura conjecture, after Japanese mathematicians Yutaka Taniyama and Goro Shimura, who developed it in the mid-1950s. A decade later, French mathematician André Weil's work on the conjecture provided the first strong evidence that it might be true. Interest in finding a proof grew following the suggestion by German mathematician Gerhard Frey in 1986 that the conjecture implied Fermat's last theorem. But, overwhelmingly, it was felt that a proof would be incredibly hard to obtain or perhaps even beyond the reach of present-day maths altogether. When, in 1995, Andrew Wiles proved the conjecture (and Fermat's last theorem) for elliptic curves of a certain class – the semi-stable variety – it caused a sensation. By 2001, the proof had been extended to include all elliptic curves and the conjecture was then renamed the modularity theorem.

Another theory that's been praised for its beauty based on forging unexpected links is monstrous moonshine, discovered and named ('moonshine' because it seemed so unlikely) by English mathematicians John Conway and Simon Norton in 1979. What's unusual, and particularly alluring, about monstrous moonshine is that it ties together two aspects of maths *and* an important area of research in theoretical physics – string theory. One of the entities with which it deals is itself considered to be among the most beautiful, bizarre, and mysterious of mathematical objects: the monster group. As its title suggests, the monster is big. It's the biggest member of what are known as sporadic simple groups, has more elements (roughly 8×10^{53}) than there are elementary particles in the entire Earth, and, disconcertingly for the imagination, exists in 196,883 dimensions.

Monstrous moonshine shows how the monster group is intimately connected with modular functions and, in particular, with a modular function known as the J-invariant or J-function. The connection was first noticed by Conway in 1978 when he was working on the monster group and modular functions separately on different days of the week. In 1992, British-American mathematician Richard Borcherds, who'd studied under Conway at Cambridge, showed how the connection arose naturally from work he'd done on a subject called vertex algebra, which, in turn, has important links with string theory.

Some people may find it hard to reconcile their notions of beauty with anything mathematical. That's not surprising given that many of us stop studying maths – usually at high school level – before it starts to get really interesting. Cambridge mathematician and professor for the public understanding of science Marcus du Sautoy has commented that the maths we learn in school is akin to scales in music – something we don't necessarily enjoy but have to know about before we can properly delve into the subject and explore it in depth. The big difference is that music is accessible through the senses so that it can give us pleasure and touch us emotionally without the need to have any abstract knowledge of it. If we're musically trained, it's true, we may be able to appreciate more of the finer points of a symphony, for instance, or a virtuoso performance, but we can all enjoy music of any genre whether we understand what goes into making it or not. At first glance, the same doesn't seem to be the case with maths. Certainly, to be able to see the beauty in an equation, a theory, or a proof you need a good mathematical background because the context and relationships with other parts of maths are all-important. On the other hand, there are some simple things in maths that can give anyone a sense of satisfaction and a taste of what it's like to experience mathematical beauty on a wider and deeper level.

Du Sautoy has pointed to a discovery made by Fermat that's easy to grasp and aesthetically pleasing. Fermat found that if a prime number when divided by four gives a remainder of one then it's the sum of two square numbers. For example, $13 \div 4 = 3$ remainder 1, and $13 = 2^2 + 3^2$; or $41 \div 4 = 10$ remainder 1, and $41 = 4^2 + 5^2$. It isn't obvious at first why this should be: why primes and squares should have such a close relationship. But if you go carefully through Fermat's proof, which isn't hard, you begin to see how the two concepts weave together and the result follows with delightful inevitability. It's taking this journey – following the unfolding of the proof – that instils the sense of aesthetic pleasure. As du Sautoy puts it: 'Like a piece of music it's not enough to play the final chord.'

In delivering school curricula, there's little time or opportunity for teachers to try to convey a sense of beauty in maths. But it isn't impossible and there are strong arguments for making the effort to develop this appreciation, even in children of primary school age. If we're touched by a subject at an emotional level, through practical engagement with it or passionate teaching, then it will stay with us throughout our lives. One approach to this kind of engagement is kinesthetic learning, in which students learn through games and activities, such as finding patterns, exploring symmetry and asymmetry, and making their own mathematical discoveries. In this way, maths and art are found to blend into one and seen to be part of a holistic whole. The stereotyping of maths as being dull, difficult, or far removed from human experience is undermined if it's presented, instead, as part of the broader, everyday world with which we're familiar. There's good reason to engage the various faculties, of touch, sight, and hearing in school maths lessons, because people process information in different ways and have different strengths. For individuals who think visually or who learn well through tactile stimulation it makes sense to use pattern blocks, cuisenaire rods (rods of various colours and lengths for understanding fractions), algebra tiles, origami, and artistic techniques of every description. In this way, students are able to grasp concepts that otherwise might elude them if they're just presented with written numbers, symbols, and formulae.

In one of the chapters of *Weirder Maths* we looked at the parallel development of maths and art, especially in areas such as projective geometry and perspective where there's been a mutual interplay and development of ideas. A number of great artists, including Albrecht Dürer, Leonardo da Vinci, M. C. Escher, and Salvador Dalí, had strong mathematical and scientific inclinations. Art can inspire maths to make progress in new fields, and maths can generate objects, such as fractals and complex geometric shapes, which, when represented visually, have great artistic merit. The beauty in these objects can be appreciated both intellectually and emotionally.

With music, it's the same. Every piece of music has a mathematical infrastructure, a fact that was first noticed by the Pythagoreans whose guiding principle was 'all is number'. Listening to, performing, or composing music, we're doing mathematics without realising it. In a way, we're able to access the beauty or other aesthetic qualities of the underlying maths indirectly, and subconsciously, through the music. It's the same when we look at any beautiful pattern or phenomenon in nature, from waves breaking on a shore to the spiral form of a galaxy. In fact, when we experience any sensation, part of that sensation involves appreciating, in some form, the mathematics that guides the ultimate cause of the sensation. Mathematicians, who've immersed themselves in the subject for many years, can experience an emotional response from the equations directly: they can appreciate mathematical beauty at source. For the rest of us, the experience comes at a high level: through physical interaction with whatever phenomenon is being ultimately guided by the maths.

Some mathematicians seem able to access the inner world of maths at multiple levels – and enjoy doing so. One of these is Canadian-American number theorist Manjul Bhargava, who won the Fields Medal – the highest annual award in maths – in 2014. Bhargava has described how he experiences the counting numbers not just in a purely abstract sense or by visualising them, conventionally, as lying on the number line. They pop up in his mind, for instance, when he sees a pile of oranges or the Sanskrit alphabet laid out in rows and columns. To him, numbers can organise themselves spatially and often appear superimposed on the real world. He also perceives them moving through time, to the rhythm of a Sanskrit poem or that of a tabla – a pair of small, Indian, barrel-shaped drums of slightly different size and shape. Bhargava is an accomplished tabla player, having learned from masters such as Zakir Hussain, and studied Sanskrit with his grandfather, who was a distinguished scholar of the language. Not surprisingly, Bhargava's perception of mathematics and the objects with which it deals overlaps with that of music and poetry. All three, he has said, have the same aim: 'to express truths about ourselves and the world around us'.

Perhaps it's no coincidence that one of us (Agnijo) is also a player of the tabla. Theoretical physicist Richard Feynman – one of the greatest scientists of the twentieth century – was an amateur drummer, too, though in his case, it was the bongos. While working at Caltech, Feynman would go down to Sunset Strip and join in with the percussion at one of the nightclubs. Many mathematicians and scientists, in fact, over the years have been drawn to music, perhaps finding pleasure from experiencing the same underlying reality – sensing the same patterns and symmetry – in what, at first, seem completely different ways. Augustus De Morgan excelled at the flute; Einstein played both piano and violin; and Hungarian mathematician János Bolyai insisted on playing violin solos after he'd defeated rival mathematicians in competitive duels. In other expressive fields, Russian mathematician Nikolai Lobachevsky, along with Bolyai one of the founders of non-Euclidean geometry, was an accomplished poet, while artists and mathematicians have a long history of collaboration and shared interests. Picasso and other Cubists, for instance, were strongly influenced by reading Henri Poincaré's popular-level book *Science and Hypothesis*, published in 1902.

Like musicians, poets, and artists, mathematicians often speak of their work as being closer to discovery than invention. The beauty of their subject, they sense, comes from outside – from some inherent balance and aesthetic necessity in the world – rather than being the product of intellectual invention. Few of us have any great talent as mathematicians or can claim it as our profession. But we can all know the beauty of nature and, through this, appreciate indirectly the elegance and grace of the mathematics that lies behind it.

CHAPTER 6

The Shape of Space

Spacetime tells matter how to move; matter tells spacetime how to curve.

– John Wheeler

THE FATE OF the universe hangs in the balance. How, or if, the universe will end depends on just one thing: its shape. Not surprisingly, finding out this shape is the holy grail of cosmology, a subject that owes as much to mathematics as it does to physics or astronomy.

In ancient times, ideas about the nature of the universe were no more than guesswork – mere expressions of which philosophy or religion you happened to favour. The Greeks, some 300 or 400 years BCE, came up, as they often did, with a smorgasbord of ideas. Aristotle, echoing the cosmic scheme of his teacher Plato, argued that Earth was at the centre of everything and that the Sun and planets moved around it on concentric crystalline spheres. All of the space beyond Earth, Aristotle believed, was filled with a fifth element with strange properties, known as the aether, and later, by medieval scholars, as quintessence. The Pythagoreans, in contrast, put an unseen central fire in the middle of the universe, with Earth, the Moon, the Sun, and the planets in motion around it, and beyond them, the fixed stars, and so came up with the first non-geocentric model of the heavens.

Opinions differed, as they still do, about whether the universe will end at some point or whether it stretches away forever in time and space. The Vedas of ancient India, alone among sacred texts, portrayed the cosmos as going through an infinite cycle of birth, life, destruction, and rebirth. Each cycle is impressively long. Carl Sagan said of Hinduism:

> It is the only religion in which time scales correspond to those of modern scientific cosmology. Its cycles run from our ordinary day and night to a day and night of Brahma, 8.64 billion years long, longer than the age of the Earth or the Sun and about half the time since the Big Bang.

The sages who wrote the Vedic texts had no way of knowing such things based on observation. It's just that their philosophy encouraged a belief in immense spans of time in the same way that the philosophy of some ancient Greeks, such as Leucippus and Democritus, led them to believe in atoms. Out of all the schools of thought in circulation back then, at least one had to be near the mark by sheer good luck!

Real progress in cosmology had to await developments in maths and physics that only started a couple of hundred years ago, together with breakthroughs in instrumentation that have allowed us to peer at objects further and further away. The idea that space itself is curved seems bizarre. After all, isn't space just emptiness – a vacuum or a void? How can empty space have a curvature?

The geometry we learn about in school would have been

familiar to Euclid, who pretty much wrote down all the rules about it in 300 BCE. Euclidean geometry was the only geometry in town until the early part of the nineteenth century. Then a few inventive minds, most notably those of Carl Gauss and another German, Ferdinand Schweikart, a lawyer by trade, started to think that perhaps there was more to the properties of shapes than had caught Euclid's eye. Nothing got published on these ideas, however, until about 1830, when, independently, Hungarian János Bolyai and Russian Nikolai Lobachevsky went public with explanations of a subject called hyperbolic geometry. Hyperboloids see their most familiar expression in the shapes of some modern structures, such as the cooling towers of power stations and, among iconic buildings, the Cathedral of Brasília and the McDonnell Planetarium at the Saint Louis Science Center. Draw a triangle on the surface of a hyperboloid, so that each

The Cathedral of Brasília, a hyperboloid structure constructed from sixteen concrete columns, each weighing 90 tons.

side (as in the case of a triangle on a flat surface) is the shortest it can possibly be connecting two corners, and a strange fact emerges: the sum of the angles is less than 180 degrees. Draw two lines side by side on a hyperboloid and, no matter how carefully you try to make them run parallel, the lines will diverge as the distance from the starting point increases.

Euclid, however, wasn't consigned to the scrapheap by these strange new results. It's just that the rules he figured out were now seen to describe a special case of all the geometries that were possible. In particular it was found that there were valid situations in which one of Euclid's axioms, or basic assumptions – the so-called parallel postulate – doesn't hold. Hyperbolic geometry was the first example to be explored of *non-Euclidean* geometry. But it was soon joined, in the 1850s, by another kind pioneered by German mathematician Bernhard Riemann. At the time, Riemann was a postgraduate at the University of Göttingen under the supervision of Carl Gauss. For an exam known as the habilitation, which candidates had to take in order to progress eventually to become a professor, Gauss suggested that Riemann write a thesis on the foundations of geometry. Riemann obliged with a masterpiece supported by a lecture, called 'On the Hypotheses Which Lie at the Foundations of Geometry', which he delivered on 10 June 1854. Twelve years later, and two years after Riemann's death at the age of only thirty-nine, the work was finally published by Richard Dedekind, although even then the scale of what Riemann had achieved was slow to be appreciated.

In Riemann's elliptic geometry, the angles of triangles add up to more than 180 degrees and lines that start off as nearly parallel as you care to make them will eventually converge. A special case of elliptic geometry is spherical geometry, illustrated by lines of longitude on Earth. For a short distance at the equator, lines of longitude are essentially parallel, but as they extend north and south they converge, eventually meeting at two points – the north and south poles.

When we learn about Euclid's geometry in school, it's always applied in either two dimensions – the plane – or three dimensions, corresponding to the everyday space we see around us. But Euclidean geometry is just as applicable in any number of dimensions, whether it be 4, 5, or 83. One of the strengths of mathematics is that it isn't limited by human imagination. The same is true of non-Euclidean geometry. We can apply it to a curved surface that's easy to visualise, such as that of a hyperboloid or a sphere. Or we can use it in situations where there are more than the three dimensions that it's possible to conjure up in the mind's eye. Non-Euclidean geometry in higher dimensions turned out to be just the maths needed to undergird some of the radically new physics that burst upon the world at the dawn of the twentieth century.

Gauss himself had been among the first to talk about the possibility of space being curved, but in muted tones. In an 1824 letter to Ferdinand Schweikart, he wrote: 'I have from time to time in jest expressed the desire that Euclidean geometry would not be correct.' He showed that just a single value was needed to describe how curved a surface was near a point in two-dimensional space – a value that became known as the Gaussian curvature. Riemann took this concept and extended it to spaces with any number of dimensions. In three-dimensional space he found it took six numbers to describe the curvature at any point, and in four-dimensional space it took twenty numbers.

Decades later, following the triumph of his general theory of relativity – a new theory of gravity – Albert Einstein acknowledged the importance of Riemann's seminal insights:

Physicists were still far removed from such a way of thinking. Only the genius of Riemann, solitary and uncomprehended, had already won its way by the middle of the last century to a new conception of space, in which space was deprived of its rigidity, and in which its power to take part in physical events was recognised as possible.

In the second part of his great lecture in 1854, Riemann posed the question: what kind of space do we actually live in? There were now choices on the geometric menu other than just plain Euclidean fare. Although the space around us appeared to obey Euclid's rules on the everyday scale of things, who was to say that over greater distances it didn't curve in a manner analogous to a hyperboloid or a sphere?

The English mathematician and philosopher William Kingdon Clifford was among the first to fully grasp the importance of Riemann's work in higher dimensional geometry. In a paper published in the *Proceedings of the Cambridge Philosophical Society* in 1876, he wrote:

> Riemann has shewn that as there are different kinds of lines and surfaces, so there are different kinds of space of three dimensions; and that we can only find out by experience to which of these kinds the space in which we live belongs. In particular, the axioms of plane geometry are true within the limits of experiment on the surface of a sheet of paper, and yet we know that the sheet is really covered with a number of small ridges and furrows, upon which (the total curvature not being zero) these axioms are not true.

Extraordinarily, Clifford suggested that matter and energy might arise from local fluctuations in the curvature of space. He also conjectured that space might be curved on a grand scale so that the entire universe was shaped like a sphere but enclosed in four dimensions rather than three. Perhaps Clifford would have gone on to pre-empt more of the breakthroughs that would eventually be credited to Einstein, but he died in 1879 from tuberculosis aged just thirty-three.

A ten-year gap separated Einstein's 1905 paper describing the special theory of relativity from his 1915 paper on the general theory. It was only during this period that he became aware of the pioneering work in non-Euclidean geometry of Riemann, Clifford, and others, which would be vital to his own efforts. Einstein's crucial insight was the so-called principle of equivalence. This recognises the fact that, for instance, if you were floating 'weightless' in a windowless room you couldn't tell (by any experiment within the room) whether you were drifting in space or free-falling under the influence of gravity. In the same way, you couldn't tell if the room were sitting still on the Earth's surface or gathering speed through space at 9.8 metres per second per second – the acceleration due to gravity. In about 1912, it dawned on Einstein that the only way to make sure that the principle of equivalence applied in all situations was to bring non-Euclidean geometry to bear.

For all his brilliance, Einstein was no outstanding mathematician. He knew about Gauss's theory of surfaces from his student days and turned to his friend and old classmate Marcel Grossman for further insights. What he discovered had a profound effect on him and his progress towards a new theory of gravity:

[I]n all my life I have not laboured nearly so hard, and I have become imbued with great respect for mathematics, the subtler part of which I had in my simple-mindedness regarded as pure luxury until now.

Grossman told Einstein of Riemann's work and the more recent breakthroughs made by Italian mathematician Gregorio Ricci-Curbastro (often referred to simply as 'Ricci') and his brilliant student Tullio Levi-Civita. The two Italians had developed a new subject known as tensor calculus. Tensors are a generalisation of vectors – quantities with size and direction – to any number of dimensions. The maths of tensors was to prove absolutely crucial to the formulation of Einstein's field equations, which lie at the heart of the general theory of relativity.

Forty years earlier, William Clifford had speculated that mass and, by extension, the force of gravity, might be an outcome of the local curvature of space. In the new world of Einsteinian physics, however, it isn't space alone that's curved but the union of space and time. This idea that the three dimensions of space and the one of time are inseparable was first proposed by French mathematician Henri Poincaré in 1905. It's also inherent in the special theory of relativity, although it wasn't until 1908 that German mathematician Hermann Minkowski, who'd once taught Einstein (and called him 'a lazy dog'!), supplied a complete geometric model of the fusion of space and time. The curvature that appears in the general theory of relativity is the curvature of Minkowski's spacetime continuum.

Forget trying to conjure up an accurate mental picture of spacetime. We think exclusively in terms of Euclidean 3D space because that's what we see around us and what

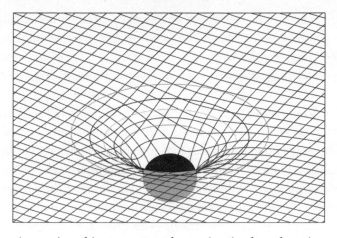

A massive object causes a depression in the otherwise almost flat surface of spacetime.

our brains are attuned to. Time seems to be something else altogether, a kind of 'other' direction in which things travel, from past to future. It's easy to grasp curvature of a 2D space: that's essentially what happens when a flat map of the world, cut into a series of curved sectors or 'gores', is folded onto the surface of a globe. A 2D space is called a surface, and we can visualise it being either flat (as in the case of a plane) or curved. A surface in an arbitrary number of dimensions goes by the more general name 'manifold'. We don't normally think of the 3D space around as having a surface, yet it does. We can only imagine this surface if it's Euclidean – the equivalent of 'flat' in 2D. It's impossible to close your eyes and see a 3D surface, or manifold, that's bent in any way. It's doubly impossible to imagine a *4D* manifold – like that of the spacetime continuum – that's curved.

But what we can imagine is irrelevant because maths transcends the power of humans to visualise. In applying the mathematics developed by Riemann, Ricci, Levi-Civita, and others, to Minkowski spacetime, Einstein showed that the curvature of the 4D surface, or manifold, in which we're embedded is determined by the distribution of matter. As an aid to our imagination – to give an inkling of the true situation – the surface of spacetime is sometimes likened to a stretched rubber sheet, such as that of a trampoline. Anything that weighs on the sheet causes it to sink down. In a similar way, the Sun, for instance, gives rise to a dip or depression in the local surface of spacetime. Anything in the vicinity of this dip will have its motion affected by the curvature. A light ray from a distant star passing close to the Sun will be deflected – its path bent slightly sunward. Likewise Earth and the other planets of the Solar System have their paths bent into ellipses, or near-circles, just as the path of a marble follows a curved trajectory when spun around the rim of a steep-sided bowl. The path followed by any free-moving object through spacetime is known as a geodesic and is the shortest way to get between two points. If the surface of spacetime is curved then the geodesic will be curved. Geodesics on Earth's surface are arcs of great circles, which explains why airliners follow such routes on long-haul flights. When depicted on a flat map, the arc of a great circle that links two cities on different continents seems far from being the shortest connecting path. On a sphere, however, it becomes clear that there's no other path more efficient in length.

Light rays, and other free-moving things such as planets and other celestial objects, will always follow geodesics through spacetime. These shortest-possible-paths are curved in the neighbourhood of a gravitating mass, noticeably so if the mass is large. Before Einstein, physicists thought only in terms of gravitational effects and the force of gravity exerted by all material things. We can still do that today: the Newtonian view of gravity works just fine for most purposes. But we now know that the 'force' of gravity is really just a manifestation of variations in the local geometry of spacetime. There's no need to scratch our heads wondering what could cause an invisible interaction to pull everything in the vicinity towards an object with mass. Thanks to the general theory of relativity, it's possible to see all occurrences of gravity as the natural behaviour of objects in moving the shortest distances possible across the curved surface of the spacetime continuum.

Anything that has mass curves the spacetime nearby. By the same token, all of the mass in the universe combined curves all of spacetime. The big question is: what's the nature of this overall cosmic curvature? There are three possibilities. The surface of the universe might be closed like that of a sphere. It might be open, stretching away forever in all directions, as in the case of a hyperboloid. Or it might be exactly flat.

In their original form, the field equations of general relativity insist that the universe can't be static – it must be either expanding or contracting. At the time the equations were first published, the prevailing view was that the universe was fixed in size. Astronomers had yet to discover that there were other galaxies beyond our own, let alone that most of these galaxies were flying away from us. To bring his theory into line with mainstream astronomical opinion, Einstein, in 1917, introduced a quantity, represented by Λ (lambda), which became known as the cosmological constant. Inserted into his field equations it exactly balanced out any tendency for the universe to change in size. He later described it as 'the biggest blunder of my life' because, by introducing a fudge factor, he missed predicting one of the great breakthroughs of the twentieth century. In the early 1930s, evidence from observational astronomers such as Edwin Hubble showed that the galaxies were speeding apart – their outward rush presumed then, as now, to be a result of the stretching fabric of spacetime.

Today, we know beyond reasonable doubt that the universe is expanding and that this expansion began about 13.8 billion years ago in an event known as the Big Bang. Until quite recently it was thought that the ultimate fate of the universe depended only on the average density of matter in space. Above a certain critical density there'd be enough matter in total to close the surface of spacetime so that, eventually, the universe would expand to a certain maximum size and then shrink back down before ending in a Big Crunch. Below the critical density the universe would be open and its contents destined to spread further and further apart, although at a gradually slowing rate over an eternal period of time. The ratio of the actual density to the critical density is a parameter known as Ω (omega).

As acceptance of the expanding universe grew, Einstein abandoned his cosmological constant. It seemed as if the only game in town now was to measure Ω and, for this, astronomers turned to their telescopes and other instruments. But it's no easy task, getting an accurate value for how much mass there is on average in a given volume of space. Account has to be taken of the fact that matter is clumped into galaxies, and galaxies into clusters, and galaxy clusters into superclusters with great voids in between. To have any chance of being near the mark, estimates have to be based on accurate models of the large-scale structure of the cosmos, which entails mapping the distribution of galaxies out to distances of billions of light-years. As well as the ordinary mass found in stars, planets, and the gas and dust between stars, factored in has to be the effective mass of photons (particles of light) and neutrinos (ghostly particles that travel at very near the speed of light). These contributions together, astronomers came to realise as their measurements improved, gave a value for Ω that fell well short of what was needed to close the universe. But then two astonishing discoveries were made.

Data hinting at the first of these breakthroughs were first collected as long ago as 1933 by Swiss-American astronomer Fritz Zwicky. He was observing the speeds at which galaxies moved within the Coma Cluster – a grouping of about a thousand galaxies that lies some 340 million light-years

The galaxy cluster MACSJ0717.5+3745, one of the most massive galaxy clusters known. It lies at a distance of about 5.4 billion light-years.

away. The speeds were too high, he found, to be explained by the total mass of the cluster. Travelling as fast as they were, the galaxies ought to break free into the inter-cluster void. For the Coma Cluster to remain bound together, Zwicky proposed, it must contain vast amounts of what he called *dunkle materie* (German for 'dark matter'). In fact, it must have a mass far in excess of that which could be accounted for by adding together all the cluster's luminous contents.

More than forty years went by before Zwicky's suggestion was revitalised. In the late 1970s, studies of the rotation of spiral galaxies made it hard to avoid the conclusion that dark matter really does exist, and in great quantities. The stars in the outer parts of the galaxies observed were orbiting around the centre much faster than could be explained by the sum total of all the bright material on display. It seemed as if there was an enormous spherical halo of unseen material in which the visible parts of each galaxy were embedded. That conclusion remains true today. Only about 17 percent of the matter in the universe is thought to be in a form that's detectable by the light, or other forms of electromagnetic radiation (radio, infrared, ultraviolet, and so on), that it gives off. The rest is dark, invisible to any of our detection equipment, and, apparently, of a completely different nature to the particles of which ordinary matter is made. Dark matter, whatever it is, bumps up the overall average density of matter in the universe. But even with the 'massive' boost from this mysterious denizen of the cosmos, the calculated value of Ω still comes out well below 1 – the figure at which the universe pivots from being infinite in size and doomed to eternal expansion to being closed and destined eventually to collapse.

If dark matter came as a shock to astronomers, it was nothing compared to the utterly unexpected discovery of another previously unsuspected component of the universe in which we live. It had been assumed that whatever the curvature of spacetime, the rate of expansion of the universe must be slowing. In Newtonian terms, the mutual pull of gravity of all the galaxies serves as a brake against the outward rush of matter that has been going on since the Big Bang. This mutual pull must inevitably, over time, it seemed, reduce the speed at which matter moved apart, irrespective of whether the expansion was eventually going to reverse or not. But then came a startling revelation.

In 1997, two teams of astronomers used the Hubble Space Telescope and ground-based telescopes to make observations of what are called Type Ia supernovae. These are explosions of white dwarfs that have acquired fresh material from a companion star until they reach a critical mass at which a runaway nuclear detonation takes place. The extreme brilliance of the explosions – briefly exceeding that of an entire galaxy – combined with the fact that the rate at which they brighten gives a measure of their true brightness, make Type Ia supernovae highly effective yardsticks for measuring the distance to remote galaxies. As well as this, how much the light from these supernovae is shifted to longer wavelengths, known as the redshift, tells us how fast their host galaxies are moving away from us. Taken together, the distance and recession speeds of Type Ia explosions serve as an accurate gauge of the rate of expansion of the universe in different cosmic eras.

The two teams of researchers were aiming to shed further light on the question of whether the universe is open or closed. In either case, they expected to find that the rate of cosmic expansion today is less than it had been in the remote past. Instead, their observations showed something astounding: the rate of expansion over time *has increased*. The galaxies are rushing apart faster today than they were billions of years ago. Evidently, gravity is not the only force at work on a cosmic scale. Something is opposing it – a kind of antigravity effect – pulling the universe apart, faster and faster as time goes on. No one knew what this totally unexpected phenomenon might be, but a name for it quickly emerged: dark energy.

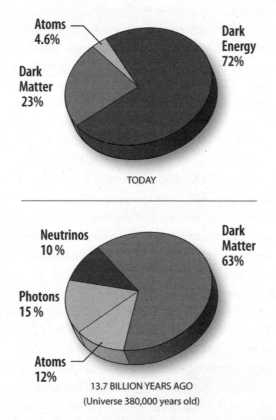

The composition of the universe today and 13.8 billion years ago.

Calculations soon revealed the extent to which we'd underestimated how much 'stuff' – total mass plus energy – there is in the universe. Based on recent observations by Planck, a European Space Agency mission, only 4.9% of the mass-energy content of the universe consists of ordinary matter. A further 26.8% is in the form of dark matter, and the rest (68.3%) is dark energy. Embarrassingly, scientists find themselves in the position of not knowing what 95% of the universe is made of.

What they have gleaned, however, is that dark energy isn't just 'out there'. It's all around us: an evenly spread (homogenous) property of the space in which we live. In a sense, it even permeates our bodies and everything we see or touch. Because it's a form of energy, it has a mass equivalence (given by Einstein's famous equation, $E = mc^2$). In these terms, its density seems extraordinarily low – about 6.9×10^{-29} kilograms per cubic metre. This is a lot less than the average density of ordinary matter in the Milky Way galaxy. But remember, dark energy exists in every corner of space with the same density, whereas ordinary matter occurs, for the most part, in relatively tiny chunks (galaxies), so that, cosmos-wide, dark energy is much more dense than the kind of mass or energy we can see.

Although scientists don't know what dark energy is, they're confident that it doesn't violate any of the known laws of physics. Its presence is allowed, for instance, by Einstein's general theory of relativity. In fact, one of the theories put forward to explain dark energy is an updated version of the cosmological constant that Einstein proposed. Unlike Einstein's famous fudge factor, however, which was introduced artificially to give a static solution to the field equations of general relativity, dark energy doesn't balance the effects of gravity. On the contrary it acts as an increasingly

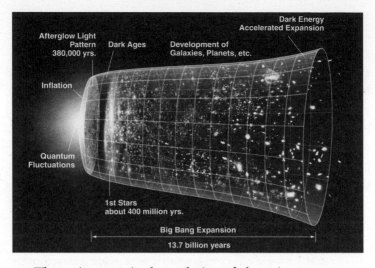

Afterglow Light
Pattern
380,000 yrs.

Dark Ages

Development of
Galaxies, Planets, etc.

Dark Energy
Accelerated Expansion

Inflation

Quantum
Fluctuations

1st Stars
about 400 million yrs.

Big Bang Expansion

13.7 billion years

The main stages in the evolution of the universe up to
the present day.

strong negative pressure that causes the universe to expand
at an increasing rate.

An alternative explanation for dark energy is that it takes
the form of a low-energy field, which came into being at the
same time as a hugely significant event that happened just
moments after the universe sparked into existence. Most cosmologists now accept that, a split second after its birth, the
universe underwent a brief but stupendously rapid period of
expansion known as inflation. Within a sliver of time lasting
from about a trillion trillion trillionth (10^{-36}) of a second to
somewhere between a billion trillion trillionth (10^{-33}) and
100 million trillion trillionth (10^{-32}) of a second, the universe
swelled in size by a factor of 100 trillion trillion trillion trillion
(10^{50}), give or take a few zeros. An important effect of this
sudden, monstrous ballooning of space and everything in it
was to magnify the tiniest of primordial fluctuations, way

smaller than the size of a proton, to large-scale dimensions. At the end of the inflationary epoch, so the theory goes, there would have been differences in the density of matter from place to place that were sufficiently large to serve as seeds for the future growth of galaxies.

At first glance, inflation and dark energy appear very different. Inflation was already ancient history by the time the universe was a microsecond old, whereas dark energy operates over timescales of billions of years. The expansion rate of the universe during inflation was 10^{50} times what it is today, whereas just 70 percent of the current rate, it's estimated, is due to dark energy. The early kind of cosmic expansion caused by inflation and the late kind caused mainly by dark energy differ a lot in detail. But they're similar in other respects. Both involve exponential expansion, albeit over vastly disparate lengths of time. Both have identical relationships, in general relativity, between the energy and the pressure to which they give rise. And for both, the relationship between time and the universe's scale is the same. These similarities have encouraged some physicists to suggest that the fields behind inflation and dark energy are tied together by something called quintessence (after the old name for a fifth classical element). The big difference between the cosmological constant idea of dark energy and the various models involving quintessence is their behaviour over time: the former is static, the latter are dynamic.

One of the great challenges of twenty-first-century physics is to determine the nature of dark energy and thereby learn the ultimate fate of the universe. Depending on which model of dark energy eventually proves correct, the universe may expand forever at an exponential rate, eventually collapse in a Big Crunch, or, in a version of quintessence known as phantom energy, expand at a rate faster than exponential and, in the remote future, end in a catastrophe known, unsettlingly, as the Big Rip. At present, observations strongly suggest that the universe will continue expanding forever. The main uncertainty is the rate. As far as scientists can tell, the expansion of the universe started to accelerate due to the effects of dark energy about 6 billion years ago. However, we're still in a relatively early cosmic phase before the expansion caused by dark energy becomes so extreme that other galaxies, or their burnt-out remains, are borne far out of sight and the material contents of space are spread out, faster and faster, into a desperately thin gruel of subatomic particles, each separated by a mind-bogglingly huge distance from its nearest neighbours.

As for the overall shape of the universe, cosmologists draw a distinction between the observable universe and the global universe. The observable universe is the part we can, in principle, see – the part from which light has had a chance to reach us since the Big Bang. It's a bubble of space, centred on Earth, that's reckoned to be about 46.5 billion light-years in radius. This is much greater than the distance of the most remote objects known – those that existed shortly after the Big Bang – because space itself has expanded in the 13.8 billion years since the universe was formed. The latest measurements, such as those by the Baryon Oscillation Spectroscopic Survey (BOSS) telescope, suggest that the observable universe is exactly, or almost exactly, flat – the 3D equivalent of a Euclidean plane, in which parallel lines never meet.

Beyond the margins of the observable universe lies the global universe – the entire extent of the spacetime that's connected with the part in which we find ourselves. On a global scale, the universe may look different from the local part we can see. On the other hand, it, too, may be flat. Even if it's geometrically flat, however, it may be folded, connected, or twisted in different ways to yield a variety of distinctive topologies. Think about a sheet of paper. Although flat, it may be folded and joined at two of its edges to make a cylinder, twisted and joined at two edges to give a Möbius band, or bent around from a cylindrical form to make a torus (doughnut shape). In the case of a universe that's flat in 3D – a so-called Euclidean 3-manifold – there are many more topological options.

In 1934, Swiss mineralogist and crystallographer Werner Nowacki proved that there are exactly eighteen different Euclidean 3-manifolds. If our universe were globally flat then, as currently seems most likely, its topology would have to correspond to one of these types. Although they're all mathematically possible, eight are thought to be physically improbable because they're non-orientable: they contain orientation-reversing loops. If the universe were one of these oddball shapes, you could in principle travel all the way around it and return with everything in your body swapped around, right for left (like Mr Plattner in the story we described in Chapter 4). More to the point, a non-orientable topology would have observational consequences that have not so far been seen. Some of the ten other shapes, such as the 1/3-twist hexagonal space and the double cube space, or Hantzsche–Wendt manifold, might also seem exotic but none can be ruled out at this stage. In fact, although the observable universe seems to be exactly flat, and there's a good chance that it continues to be flat beyond where we can see, the global universe might have a hyperbolic or a spherical curvature, vastly increasing the range of possible topologies.

A favourite question of adults and children alike is: what's outside the universe? The simple but not very satisfying answer is that there doesn't have to be an outside. Just because we can't imagine a space that doesn't have anything surrounding it and anything to expand into, doesn't mean that it's impossible. It's perfectly feasible, both mathematically and physically, that all the space there is might be contained within the universe. According to some theories, including that of the multiverse, there may be other regions of spacetime separate from our own. But, even if these 'other universes' exist they're independent from our own and not some larger space in which our own is embedded.

CHAPTER 7

Life by Numbers

Everything is what it is because it got that way.

– D'Arcy Wentworth Thompson

MATHS PERMEATES LIVING things – from the electrical activity of our hearts to the shape of shells and the flight of birds. It's woven into the fabric of life, from the molecular level – the form and structure of biomolecules – to the forces that govern the behaviour of creatures en masse, such as the murmurations of starlings, or the waves of extinction and proliferation that affect whole species.

Some of the most complex and unexpected maths in biology involves, oddly enough, the simplest of life forms – organisms so small and elementary that they lie on the margins between life and non-life. The shape of viruses, it turns out, is best understood in terms of higher dimensional geometry. Whereas the classical geometry of Euclid is formulated in 2D (the plane) or 3D (ordinary space), the lattice-like arrangement of genomic material on the surface of viruses leads to something far more unfamiliar: geometry in six dimensions. This isn't to suggest that viruses are actually six-dimensional but rather that the maths of six dimensions is a good way to understand the shape of viruses in 3D. The convoluted three-dimensional forms into which viruses are arranged have been found to be the equivalent of shadows or slices of simpler, six-dimensional objects.

The mathematical patterns of life may be specific to a particular organism or class of organisms, or they may be broad and sweeping. In the 1930s, Swiss agricultural biologist Max Kleiber made a surprising discovery about the way a creature's basal metabolic rate (BMR) – the amount of energy it uses while at rest – is related to its body mass. Before analysing data from various studies that had been carried out, Kleiber had assumed that BMR would vary as body mass to the power 2/3. Like others, he'd supposed that an animal's metabolic rate would be in proportion to its need to get rid of excess heat, so that it would follow what's known as the square-cube law. Increase a body in size by a certain linear scale factor, this law says, and its new surface area will be proportional to the square of that scale factor while its new volume will go up by the cube of that factor. What Kleiber found, however, was something intriguingly different: an animal's BMR scales, not to the two-thirds power of its mass, but to the *three-quarters* power. 'Kleiber's law' also applies to both heart rate and lifespan. In general, the more massive an organism, the slower, on average, its heart rate, and the longer it lives, with heart rate decreasing and lifespan increasing as three-quarters the power of the mass. Humans, as in so many other ways, are exceptional. Our average lifespan is somewhat greater than that of an African elephant despite the fact that we weigh fifty times less. Broadly speaking, though, Kleiber's law holds true across a huge mass range of animals from bacteria, weighing about a trillionth of a gram (10^{-12} g), to blue whales, weighing up to 100 tons

(10^{11} g) – a span of twenty-three orders of magnitude. It also applies, remarkably, to plants.

The 3/4-power law has been confirmed time and time again by studies over the years. Yet it was only in 1997 that a team of physicists and biologists came up with a credible explanation for it. Their theory involves fractals – structures in which similar patterns recur repeatedly at smaller and smaller levels. It takes into account the fact that there's a major problem in supplying the tissues of larger organisms. To carry oxygen and other materials to all the cells in our bodies we depend on elaborate distribution networks. We need a complex circulatory system, for instance, and an intricate network of passageways in our lungs. Smaller organisms don't face the same difficulty. Single-celled and simple multicellular creatures have such big surface-area-to-volume ratios that they can get all the oxygen they need directly through their body walls. We, on the other hand, have to pipe in all the essentials for life through a many-branched, hierarchical system of channels of bewildering intricacy – as do other large animals and plants (the internal passageways of which carry water and nutrients from the soil). So effective are these fractal-like structures at maximising surface area that their total area occupies space nearly as much as if it had an extra dimension. The mathematical upshot is the extra dimensionality of the power law describing their effect on metabolism – not 2/3 but 3/4.

In life, efficiency is everything. To have any hope of surviving in the wild long enough to pass on your genes to the next generation, you can't afford to waste energy or be less effective than your rivals at exploiting whatever resources are to hand. It's not surprising then that living things are masters of making the most of whatever's important to them. Plants need to maximise the amount of light energy they harvest with their leaves. If the leaves are arranged around a central stem, it's not much use if those in front shade the ones behind. Instead, each successive leaf along the stem must project at a different angle from the one ahead of it. Any kind of periodic arrangement, such as would result by rotation through a rational fraction of a circle (1/3, 1/4, 2/5, and so on) would be bad because light to lower leaves would be completely blocked by some of those higher up. The only way to ensure there are no exact overlaps is for each successive leaf to be rotated from its predecessor through a fraction of a whole turn – the so-called phyllotactic ratio – that's an *irrational* number. A number is irrational if it can't be expressed in the form of one whole number divided by another.

Phyllotaxis in *Aloe polyphylla.*

You might expect that if irrational phyllotactic ratios are better for effective light-capture, then the best of all would be the one that's *most* irrational. And, sure enough, this is what's found in nature. An irrational number can be approximated by a sequence of rational numbers. For example, the sequence 3, 22/7, 333/106, ... gives a steadily improving approximation of pi. If you want to approximate pi to the nth decimal place, there's a fraction in this sequence that will be accurate up to that point. Different sequences converge at different rates. If you want the value of pi to fifty decimal places, the sequence 3, 3.1, 3.14, 3.141, ... will deliver that level of accuracy by the 51st member. However, the sequence 3, 22/7, 333/106, ... will get you there much quicker. For any irrational number, it's possible to measure the speed of convergence of the most optimal sequence of rationals that approximates it.

Of all irrationals – and there are infinitely many of them – the one that has the slowest converging optimal sequence is ϕ (phi), the golden ratio. Phi has the value 1.61803... and the most commonly found phyllotactic ratio in the plant world is $1 - 1/\phi = 0.382$. To make a more direct comparison you could say that in such a plant each new leaf is rotated 1.618..., or ϕ, turns from the previous one – in other words, a full turn plus 0.618... of a turn (360 degrees plus 222.5 degrees). But it makes more sense to omit the full turn, leaving 0.618 (equal to $1/\phi$), and then, to further simplify, think of this as a rotation in the opposite direction, yielding 0.382 of a turn, or 137.5 degrees.

The golden ratio is intimately connected with a sequence of numbers that was introduced to the West by Italian mathematician Fibonacci (aka Leonardo of Pisa) in his 1202 book *Liber Abaci* (The Book of Calculations). The Fibonacci sequence is 1, 1, 2, 3, 5, 8, 13, 21... each Fibonacci number being equal to the sum of the previous two. Successive ratios of Fibonacci numbers give a steadily improving approximation to ϕ: $1/1 = 1$, $2/1 = 2$, $3/2 = 1.5$, $5/3 = 1.66...$, $8/5 = 1.6$, $13/8 = 1.625$, $21/13 = 1.615...$

To illustrate how his sequence builds up, Fibonacci used the example of a growing population of rabbits. In doing so, he effectively launched the subject of theoretical biology – the application of maths to the life sciences. In *Liber Abaci*, he wondered how many pairs of rabbits would be produced starting with a single pair under ideal circumstances. The assumptions he made were: no rabbits die or are eaten by predators; every female reproduces every month, starting from the second month after she was born, and every birth yields one male and one female rabbit. At the end of each month, the numbers of pairs of rabbits are the numbers in the Fibonacci sequence.

The seed head of a sunflower.

Fibonacci numbers are often in evidence in the arrangement of seeds on flower heads. Look closely at a sunflower and you'll notice that its seeds lie along two spirals that wind in opposite directions out from the centre. Each seed sits at the very angle from its nearest neighbours – 137.5 degrees – that's also most prevalent among leaves. Arranged in this way, the number of seeds in each turn of the spiral follows the Fibonacci sequence: 1, 2, 3, 5, 8, 13, 21, 34, 55, 89… Whereas with leaves the golden angle of 137.5 degrees maximises the amount of light that can be collected, in the case of seeds it optimises their packing. Even slight variations from the golden angle, as seemingly trivial as one-tenth of a degree, quickly disrupt the tight-knit way in which the seeds are arranged.

Plants don't have to calculate where they're going to place their leaves or seeds as they grow. In the same way, when a mollusc, such as a snail or a nautilus, enlarges its shell, it isn't mentally trying to figure out the best way to lay down the next layer of chalky exoskeleton. Life forms, and inanimate things (like ocean waves and sand dunes) come to that, are naturally mathematical because maths – and every aspect of science, which is based on it – infuses the universe at every level. We're all inherently mathematical beings, whether we think we're 'good' at maths or not.

The Fibonacci spiral that governs the placement of leaves and seeds in many plants is an approximation of the golden spiral, which gets wider by a factor of ϕ for every complete rotation it makes. Golden spirals, as we've seen, have properties that give their owners advantages in the survival stakes so that evolution has selected for them over many millions of years. They're mathematical and physical facts made manifest in the makeup of living things. Golden spirals, in turn, are a particular form of a spiral known as a logarithmic spiral.

Nautilus half-shell showing the chambers arranged in a logarithmic section.

Logarithmic spirals are a common sight in the natural world because of two important characteristics: self-similarity and equiangularity. Self-similarity means that they look the same at every scale of magnification. Equiangularity refers to the fact that, in a logarithmic spiral, the angle any tangent to the curve makes with a tangent to a circle at the same radius – the pitch angle – is constant. In the case of a golden spiral the pitch angle is about 17.03 degrees. In other logarithmic spirals, the pitch angle may differ but is always constant for a given spiral. Any process that turns or twists at a constant rate while moving with constant acceleration will generate a logarithmic spiral. The clearest and most exquisite example is the internal structure of the shell of the nautilus, revealed when such a shell is sliced in half. As it grows the animal relocates from one chamber to the next, which it constructs according to a single, unchanging blueprint – a mathematical plan encoded in its genes at the moment of conception.

The logarithmic spiral occurs not just in the shape of organisms – for example, the way a chameleon curls its tail – but also in the way some animals behave. It's seen in the flight pattern of peregrine falcons, most probably due to the equiangular property. When diving at small prey from a great distance (up to 1,500 metres away) and at high speed, a peregrine has a dilemma. While flying straight ahead, it achieves maximum visual acuity with its head cocked 40 degrees to one side so that the image of the prey falls on the most sensitive part of its retina (the deep fovea). But with its head in this position, it experiences more drag and so is slowed down, giving its intended victim time to escape. By approaching along a logarithmic spiral the falcon can keep its head straight while homing in at top speed with the prey in sharp view.

After Fibonacci, there were few new developments in the biological application of maths until the eighteenth century. In 1760, Swiss mathematician and physicist Daniel Bernoulli submitted a paper to the Academy of Sciences in Paris. At the time, the practice of inoculation was still relatively new and controversial. Bernoulli's paper argued that although vaccinating against smallpox might lead to a few extra deaths, the long-term benefits far outweighed the immediate risks. Some thirty years later, English scholar and cleric Thomas Malthus wrote his famous book on human population growth. *An Essay on the Principle of Population* predicted a bleak future in which, without some constraint on birth rate, human numbers would double every quarter of a century while food production lagged behind, resulting in widespread famine and starvation.

At the dawn of the twentieth century, German botanist and philosopher Johannes Reinke became the first to use the term 'theoretical biology'. It was a subject explored in great depth by Scottish biologist and mathematician D'Arcy Thompson in his 1917 book *On Growth and Form*. For thirty-two years, Thompson served as Professor of Natural History at University College, Dundee – the city in which both the authors live – before moving to nearby St Andrews for a similar period of time. He's best remembered for his studies of morphogenesis – the emergence of structures and patterns in living things. Thompson's writings on the mathematical basis of biological forms and on mathematical beauty in nature later struck a chord with other influential thinkers, including evolutionary biologist Julian Huxley, anthropologist Claude Lévi-Strauss, architect Le Corbusier, and mathematician Alan Turing.

In 1952, Turing wrote an article called 'The Chemical Basis of Morphogenesis', which was to be his only published work in biology. In it he asked: how can a small, uniform ball of cells – the embryo in its most primitive state – develop into an increasingly complex structure with, eventually, all the distin-guishing features of a spider, a seahorse, or a human being? How is the initial perfect symmetry broken merely by the action of molecules randomly moving around and reacting? Turing theorised that among the ingredients of the earliest stage of the embryo are molecular components he called mor-phogens ('shape-formers'). These are responsible, he argued, for the controlled introduction of asymmetry, including the appearance of distinct structures and tissues. His paper may have attracted more attention during his lifetime had it not been eclipsed by the discovery of the double-helix nature of DNA, announced by Francis Crick and James Watson just a year later. As it was, decades passed before biologists began to look more closely at Turing's ideas about the emergence of natural patterns from an initially uniform state.

Turing's model is, in essence, very simple. It calls upon just two interacting agents, an activator and an inhibitor, which diffuse through the substance of the primitive embryo like drops of ink in water. The activator starts off some process, say the formation of a stripe, while also causing more of itself to be produced. Unchecked, the activator would soon run riot, covering the emergent organism in one giant stripe. The inhibitor, however, as its name suggests, curbs the activator by diffusing faster and stopping it in its tracks. The two chemicals, effectively a creator and a destroyer, work in tandem, resulting in localised regions of activation and the appearance of a series of stripes, or, as the case may be, dots or other patterns.

Recently, researchers at the University of Florida found that dermal denticles – toothlike projections – in the skin of sharks are laid down in accordance with Turing's morphogenetic mechanism. In fact, it turns out, shark denticles are generated by the same genes and the same method that, in birds, control the pattern of feather formation. This new work suggests that Turing's patterning process may have been at work from the earliest days of vertebrate evolution, directing the formation of features in the developing embryos of a wide range of backboned species.

It's clear that you don't need much of a brain, or a brain at all, to be able to do maths. Evolution, and the battle to survive, supply the necessary built-in savvy without having to crunch numbers, from one minute to the next, in a giant, energy-hungry assemblage of neurons. Army ants are master builders, even though their brains are minuscule and they're practically blind. Lacking any permanent home, they march through the jungle in columns up to 100 metres long, containing hundreds of thousands of individuals, on foraging expeditions known as raids. They've no leader or chief architect to tell them what to do if they meet an obstacle. Yet often they come to gaps in their path and somehow have to span these or find a way around them.

Suppose you happen to be the ant at the head of a column of marauding comrades and you come to a gap. Instinct tells you to stop in your tracks, but as soon as you do so your formicidean chums start to trample over your back (since you're now part of the path). Unlike in human society, you don't take offence or panic at being trodden under foot but instead, motivated by your genetic programming, you freeze in place. The first ant to cross over you does likewise when they reach the edge of the gap, which has been shortened slightly by the length of your body. So the process continues, individuals locking legs, until the gap has been spanned by a living bridge across which the rest of the colony can march to the other side.

But there are complications. Say your raiding phalanx comes to a V-shaped gap at its widest point. This is like the human problem of deciding how best to build a road across a tidal estuary when the most direct route connecting roads on either side is at the river's mouth. You might choose to build a very long bridge there to minimise the distance that needs to be travelled but this would involve a hugely expensive construction project. Alternatively, you might opt for a short bridge well upstream, thereby keeping the cost down but greatly lengthening the journey. In the case of the ants there's a further problem. On any given march they might have to negotiate several dozen gaps at the same time to accommodate the whole length of the column. But ants that form part of temporary bridges aren't available for foraging, so there has to be some trade-off between building distance-minimising bridges, which occupies a lot of ants, and leaving enough individuals free to do the job of finding and transporting food. Researchers have found that army ants effectively use an algorithm to carry out the necessary cost-benefit analysis.

When a significant fraction – around a fifth – of the marching column is tied up in spanning gaps, the ants change their behaviour. They stop making the longest bridges (that is, taking the shortest routes) possible. They also start to dismantle some existing bridges. An ant that's temporarily occupied as part of a bridge doesn't know, of course, what's happening in the rest of the column but it doesn't have to. Researchers have found that an ant is sensitive to the amount of traffic passing over its back. As long as the traffic flow is high it stays locked in place, but when the numbers marching overhead drop below a certain threshold, perhaps because too many other ants are parts of bridges themselves, the ant unlocks itself and rejoins the march.

Over many millions of years, natural selection has given the army ant this ability to solve on the fly what, at first glance, looks like a complex resource allocation problem. In unravelling the secrets of different kinds of animal behaviour, biologists often find mathematics at work. And, increasingly, mathematicians find themselves attracted to research in biology.

As a teenager, Corina Tarnita (like Agnijo) had enjoyed success in maths competitions. In three successive years, from 1999 to 2001, she took first prize in Romania's National Mathematical Olympiad and was then offered a place at Harvard. Having earned her bachelor's degree, she progressed to Harvard's graduate school to study aspects of pure maths. But then something happened: she found her interest waning. Abstract problems in maths no longer held her interest as they had and she started to consider careers that had more to do with the real world. Around this time she came across a book in the university library called *Evolutionary Dynamics: Exploring the Equations of Life* by mathematical biologist Martin Nowak, which captured her imagination. By good fortune, Nowak was a professor at Harvard, so Tarnita emailed him and they arranged to meet. Shortly after, Tarnita switched to Nowak as her PhD supervisor. The two of them, in collaboration with the renowned biologist Edward O. Wilson, began a research project into the evolution of cooperative insects, including ants and termites, which culminated in a 2010 paper in the journal *Nature*. Since then, Tarnita has continued her work into how living things orchestrate themselves into patterns on different scales, from the individual to the collective.

Jessica Flack, another transdisciplinary scientist who does research into living ensembles, co-runs the Collective Computation Group at the Santa Fe Institute, New Mexico. But whereas Tarnita studies collective behaviour of animals in time and space, Flack also looks at the social dimension. One of her early projects focused on a group of macaques, the stability of which was maintained by a few tough-looking males who would break up any fights that started between less dominant monkeys. Although the interventions of these macaque 'police' took place at an individual level, they produced a ripple effect, like the collective movements of a flock of birds, on the whole troupe. Commenting on macaque society, Flack said: '[T]heir metric space is a social coordinate space. It's not Euclidean.' Today her group at Santa Fe studies collectives as diverse as slime moulds, neurons, and even the Internet, trying to understand the general rules that may orchestrate them all.

Developmental biologist Ann Ramsdell had a more personal reason for applying maths to the life sciences. In 2009, she was diagnosed with stage 3 breast cancer (in which the cancer has spread to nearby lymph nodes but not yet metastasised). Searching the scientific literature to learn more about her chances of beating the disease, she made a surprising discovery. The odds of recovery were different for women who had cancer in the right breast than in the left. What's more, the chances of developing cancer in the first place were greater in women with asymmetric breast tissue.

Years earlier, while defending her dissertation, Ramsdell used a borrowed slide of a chick embryo that showed heart looping. This is the stage at which the initially straight heart tube begins to form a more complex structure reminiscent of the adult heart. The problem was she'd put the slide in backwards, as was later pointed out to her by a colleague. She'd never given much thought to the fact that the developing heart of a chicken 'knows' right from left, as does our own. Now it became her main point of focus, and for her postdoctoral research she studied why the heart loops preferentially to one side.

Fortunately, Ramsdell made a complete recovery from her cancer. But the experience motivated her to turn away from the heart and study instead asymmetry in the mammary glands of mammals. Her work began to focus on different types of cells in the breast and the genes and proteins that are active within them. The left breast, which is 5 to 10 percent more prone to cancer than the right, also has more unspecialised cells. These cells give the breast a greater capacity to repair damaged tissue but, because they can divide faster, are also more likely to be involved in the development of tumours. To understand why unspecialised cells favour the left breast, Ramsdell and her colleagues are now looking at asymmetry in the embryonic environment in which the cells first appear.

Scientists are finding mathematics at work in every aspect of living organisms and are applying maths more and more to biological problems. But they're also finding that various kinds of mammals and birds, and even some insects, have numerical skills of their own. A basic arithmetic ability may help an animal in defence, gathering or hunting for food, or in reproduction. Serengeti lions, for instance, are good at judging the size of rival prides. From the roars made by another group, they can tell if they're outnumbered or not, and will only attack or defend if the odds are in their favour. Chimpanzees and various species of monkey also have this talent to weigh the strength of the opposition based on sound, and hyenas have been found to be especially adept.

Many animals can tell at a glance which of two or more sources of food is more plentiful. The survival advantage of this is obvious, so there's evolutionary pressure that selects for the ability to assess relative quantities accurately. But being able to recognise that one quantity is larger than another isn't the same as being able to count. True counting means having an intuitive notion of ordinality: the idea that there's a sequence, where one is followed by two, two by three, and so on. Frogs, it turns out, have this knack. Female frogs can identify male members of their species by the number of pulses in their croak. Researchers have found that they can do this for phrases up to ten notes long.

While frogs count to help them mate, bees do so for the purpose of navigation. A worker honeybee flies from its hive looking for food and, when it's successful, collects some and returns home. By placing landmarks and training bees to receive a food reward after they'd passed a specific number of them, scientists found that the bees can keep track of the number of landmarks. What's more, they can count the landmarks even when the objects used are changed, so that, like us, bees are capable of object-independent counting.

Members of the crow family are even better at this kind of numeracy. Shown a display containing a certain number of objects, they can then find a display of different objects containing the same quantity. New Zealand robins get visibly upset when, in tests, they're cheated out of a promised number of mealworms. Meanwhile, Alex the famed African grey parrot could accurately add together three sets of objects and numerals to the sum of eight, putting him on a par with the maths ability of chimpanzees and the average five-year-old child. Of all non-human species, however, dolphins may be the outstanding mathematicians. In their use of echolocation pulses and bubble rings to confuse and capture prey, dolphins appear to be using complex nonlinear maths that goes beyond the capabilities of other animals – even most humans.

CHAPTER 8

Stats Weird

Facts are stubborn things, but statistics are pliable.

– Mark Twain

You're more likely to die from a falling coconut than from a shark attack, and more likely to die on your birthday than any other day of the year. The average person falls asleep in seven minutes and, over a lifetime, spends twenty-five years sleeping. About 11 percent of the population are left-handed. The most typical human face on Earth is that of a 28-year-old Chinese man.

These are statistics. They're not statements that apply to anyone in particular. Take one of us (David), for instance. As I live in Britain and rarely swim in the sea, the prospect of death by falling coconut or shark bite isn't something that keeps me awake at night. What *does* keep me awake are ideas for the next day's writing: drifting away inside seven minutes is something I can only dream about. And, I'm neither Chinese nor, unfortunately, twenty-eight.

It's easy to confuse the terms 'statistics', 'data', and 'facts', which leads to them being used interchangeably. It isn't a statistic, for example, that water expands by nine percent when it freezes: it's a fact, guaranteed by the laws of nature. On the other hand, the statement 'your foot is the same length as your forearm' is based on statistics and certainly doesn't apply to everyone. That the Eiffel Tower has 1,792 steps is a fact, that the average hen lays 228 eggs a year is a statistic. Statisticians work with aggregated data and come up with conclusions from a series of observations or measurements, not individual items. These conclusions may be very useful or extremely misleading, depending on how the data were collected and analysed.

The expression 'There are three kinds of lies: lies, damned lies, and statistics' was popularised by Mark Twain and attributed by him, in his autobiography, to the nineteenth-century British Prime Minister Benjamin Disraeli. But Twain's assertion itself is untrue because Disraeli had nothing to do with it. In fact, the origin of the phrase is uncertain, although variations on it appeared in print throughout the 1880s and 1890s. The *gist*, of course, holds true: that the misuse of statistics, whether accidental or deliberate, can lead to all sorts of false conclusions and crazy beliefs. Sometimes errors get repeated so often that completely wrong data are taken to be facts without anyone bothering to check them.

Statistics is fertile ground for mix-ups, through oversight or intent, which skew predictions and, in the worst cases, totally misrepresent the reality of a situation. Cherry-picking data is a classic ploy of politicians or anyone else trying to bolster their opinions or pet theories and, in elections, pollsters often fall victim to a combination of errors. In the 1948 US presidential election, three major polls predicted that Thomas Dewey, governor of New York, would win against the incumbent, Harry Truman – and got it wrong. Among their mistakes, they stopped polling too soon and so failed to take

Harry Truman holds up a newspaper with an erroneous headline claiming that Dewey had won the 1948 presidential election.

into account the late energising effect that Truman tended to have on voters. Telephone polls also favoured Dewey because at that time only more affluent people tended to have phones – and the well-off were more likely to be Dewey supporters.

Samples are the lifeblood of statistics. They supply the raw data with which statisticians work. If the sample is chosen poorly – if it's biased and doesn't accurately represent the population as a whole – then the results of the statistical analysis will be flawed and misleading.

Like probability, a subject to which it's closely allied, statistics can fly in the face of what intuition tells us. Take a case where you have the choice of going to one of two hospitals, A or B. You've unfortunately contracted Bloggs syndrome (BS), which can be deadly, and are trying to decide where to go to have the best chance of survival. Of the last 1,000 people with BS treated at Hospital A, 550 have lived to tell the tale – a survival rate of 55 percent. Of the same number with BS who were admitted to Hospital B, only 300 emerged alive, for a survival rate of 30 percent. The choice seems obvious. You're about to head off to Hospital A when a friend, who happens to know a thing or two about statistics, points out a complication. BS comes in two forms, mild and severe, and you don't yet know which form you have. She shows you the breakdown of the survival rates for each hospital, recently published in the *Journal of BS*.

	Hospital A survival rate	Hospital B survival rate
Mild Bloggs syndrome	534 out of 860 (62%)	163 out of 213 (77%)
Severe Bloggs syndrome	16 out of 140 (11%)	137 out of 787 (17%)

It turns out that, regardless of whether you have the mild form or the severe form of Bloggs, you're better off going to Hospital B! You'd been misled because when the separate figures are combined, Hospital A seems better. The reason for this is that Hospital B admits many more patients with severe BS than Hospital A and, even though these patients are more likely to survive in Hospital B, they still have a much lower survival rate than patients with only the mild form.

This effect, in which a trend that's evident when groups are considered separately but seems to disappear or even be reversed when data from the groups are combined, is known as Simpson's paradox. It's named after Edward H. Simpson, a British statistician, civil servant, and former wartime code-cracker at Bletchley Park, who first described it in 1951. The paradox pops up in different guises but all have one thing in common: connections between data that show up clearly when the data are looked at in separate groups are lost or seem to be turned around when the data are amalgamated.

In 1973, officials at the University of California at Berkeley were worried. The graduate division had admitted 44 percent of male applicants but only 35 percent of female applicants. Concerned that the school might be sued for gender bias, the associate dean asked Berkeley statistician Peter Bickel to delve into the issue and see what was going on. With all the applications lumped together there was indeed a larger fraction of men admitted than women. But a very different story emerged when Bickel and his team considered each department separately. It turned out that women tended to apply to graduate departments, such as social sciences, that were more difficult to enter for applicants *of either sex*. When this fact was taken into account, there was actually a small but statistically significant bias in favour of women. Interestingly, Bickel's analysis did point to discrimination against women but at earlier stages in their education. The graduate departments at Berkeley that were easier to get into, but that women tended to avoid, were those that required students to have done more maths at undergraduate level. Bickel and his team concluded:

> The bias in the aggregated data stems not from any pattern of discrimination on the part of the admissions committees…but apparently from prior screening at earlier levels of the educational system. Women are shunted by their socialization and education towards fields of graduate study that are generally more crowded, less productive of completed degrees, and less well funded, and that frequently offer poorer professional employment prospects.

Simpson's paradox speaks to the fact that in statistics the connections between things can seem to shift depending on how the data are handled or grouped. It's a good reminder to us all, and researchers in particular, not to jump to conclusions about cause-and-effect relationships.

That most prolific spinner of perplexing aphorisms, baseball legend Yogi Berra, once said: 'You can see a lot by looking.' Nowhere is that more true than in statistics. In 1973, the English statistician Francis Anscombe came up with a striking example of how sets of data with more or less the same statistical properties, such as mean (average value) and variance (how much the data are spread out from the average), could look totally different when graphed. He'd become interested in statistical computing and was keen to stress that 'a computer should make both calculations and graphs.'

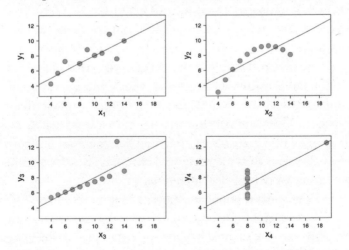

The four data sets defined by Francis Anscombe for which some of the usual statistical properties (mean, variance, correlation, and regression line) are the same, even though the data sets are different.

Anscombe wrote a paper in which he showed four graphs that became known as Anscombe's quartet. The sets of eleven data points in each of the graphs were almost identical in terms of their basic summary statistics, including mean, variance, and correlation. Anyone looking at the numbers alone would have concluded that when plotted out on squared paper, with the usual x and y axes, the sets of points would have looked quite similar. In fact, it was commonly assumed among statisticians that once the numbers had been crunched, graphing was mostly irrelevant. Anscombe showed how off the mark that assumption could be. The four sets of data he constructed, despite having the same basic stats, looked totally different when plotted out. The first yielded points scattered at various distances from a straight line of best fit. The second produced a smoothly arcing curve that was very obviously nonlinear. The third and fourth data sets also gave rise to highly patterned arrangements of the points, with the exception, in each case, of a single renegade outlier.

Statistics can easily fool us if used incorrectly, or if we fail to take in the whole picture of what's going on. The situation is even worse when data are presented in a way that's deliberately misleading – as often happens in advertising and politics. Without resorting to outright lies, there are plenty of ways to distort data to create a false impression. A common ploy is the use of the specious chart.

Bar charts are a popular means of representing, and misrepresenting, data. Suppose there are two brands of toothpaste: Brand A has sold 10.7 million tubes over the past six months whereas Brand B has sold 10.2 million. Comparing the numbers, there's obviously not much to choose between them. But Brand A's advertising department has come up with a bar chart that makes the difference look huge.

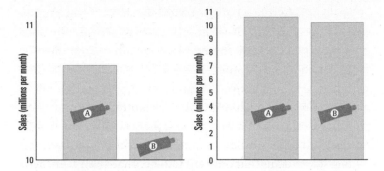

Two bar charts showing the same information but giving different impressions. On the left, sales of two brands of toothpaste are shown but with the vertical scale starting at 10 million. In the chart on the right, the scale starts at zero and the difference in sales of the two brands looks dramatically less.

A quick glance at the chart, which is all that many potential customers will give, especially if it's flashed up for a few seconds in a TV ad, suggests that Brand A is more than three times as popular as Brand B. But looking more closely we can see that the vertical axis, showing sales, doesn't start at 0 but instead at 10 million. While in reality the difference in sales is only slight, the chart makes it appear that Brand A has far outsold its competitor.

Pie charts come in for a similar kind of manipulation. People generally find it harder to compare angles than lengths. Even an ordinary pie chart can lead to some distortion in the mind of the viewer, especially when the slices aren't labelled with actual values. But the distortion is much greater in the case of three-dimensional pie charts, which appear cylindrical, like roundels of cheese, and in which any slice near the front looks much bigger than a slice at the back. A similar deception can be used in pictograms, when individual pictures differ in size, making whichever category has the largest pictures look dominant, or, when comparing two different values using a 3D figure. In the latter case it may not be clear whether values are being represented by the width of the figure or its volume – a factor that has a huge effect on perception.

Unless we're careful we can end up reaching all kinds of faulty conclusions because of a failure to interpret properly the underlying statistics. Some of these problems can involve literally a matter of life and death, as in the case of the so-called 'prosecutor's fallacy'. To understand what this is about, suppose an airport has installed a security scanner designed to detect armed terrorists. It has a 99% success rate. In other words, if a terrorist passes through, there's a 99% probability that they'll be reported as such, whereas the scanner incorrectly flags innocent folk as bad guys only 1% of the time. Now, suppose that someone passes through the scanner and the alarm goes off. What's the probability that they're actually a terrorist? It's tempting to think the answer is 99%. But this conclusion ignores the fact that the chances a person chosen at random is a terrorist is extremely low. Suppose, for the sake of argument, that one in a hundred thousand people who enter the airport is a terrorist. Then, out of 10 million people, on average one hundred are terrorists and, on average, 99 of them are correctly identified as such by the scanner. The other 9,999,900 people who pass through are innocent yet our scanner will mistakenly identify 99,999 of them as terrorists. The probability that someone flagged by the system is actually a terrorist is a mere 0.1%.

An instance of the prosecutor's fallacy in recent legal history involved a British woman, Sally Clark. Both of Sally's children died of what appeared to be sudden infant death syndrome (SIDS), otherwise known as cot death. This coincidence of events seemed so suspicious that, in 1999, she was put on trial for murder. The prosecutor argued, using statistical evidence presented by paediatrician Roy Meadow, that the chance of one child dying from cot death was about 1 in 8,500 so that the chance of two siblings dying this way was 1 in 8,500 squared, or 1 in 73 million. These odds were so low, the prosecutor concluded, that it was much more likely that the children had been killed by their mother.

Sally Clark was handed a double life sentence for murder. But it wasn't long before flaws began to emerge in the statistical analysis. In 2001, the Royal Statistical Society issued a press release in which it stated:

> The case of R v. Sally Clark is one example of a medical expert witness making a serious statistical error, one which may have had a profound effect on the outcome of the case.

The following year, Ray Hill, a professor of mathematics at Salford University, laid bare the problem with the prosecution's case. A proper examination of the statistical data on SIDS revealed that if one child in a family dies from cot death then the probability of a sibling suffering the same fate is much greater than 1 in 8,500 because of genetic factors. The crucial point in the Clark case, though, is that the probability which matters is not the probability that both children will die of cot death given that Clark was innocent, but rather the probability that Clark was innocent given that both children died. Double cot death is very rare but double infanticide by a mother is even rarer – by a factor, Hill estimated, of between 4.5 and 9. Therefore, a priori, before any other evidence was considered, Clark would be much more likely to be innocent than guilty. In fact, it later emerged that an autopsy of her second son to die, eight-week-old Harry, had shown that he was suffering from a staphylococcal infection that may have been the cause of his death – evidence that was not disclosed by Home Office pathologist Alan Williams at the trial.

In 2003, Clark's conviction was overturned, along with those of two other women who'd been imprisoned for murder of their babies on the basis of similar faulty statistical evidence supplied by Meadow. The paediatrician was subsequently struck off by the General Medical Council but later reinstated on appeal. Williams was removed from the register of Home Office pathologists but also later reinstated. Unsurprisingly, Sally Clark, who had been a practising solicitor, struggled with her psychological health while in prison and, upon release, never recovered from the trauma to which she'd been subjected. She died in 2007 from acute alcohol poisoning.

Just like a clever magic trick or an optical illusion, problems in statistics and probability can expose shortcomings in our common-sense perceptions. Take the case of Steve, for instance, who's shy and withdrawn, and has a need for order. Is he more likely to be a librarian or a farmer? Instinct may tell you 'librarian' but when you bear in mind how few librarians there are compared with farmers it becomes clear that Steve, whatever his personality type, probably works with the land rather than with books. In the same way, if you're asked if a heavily tattooed young person with long black hair and a penchant for death metal is more likely to be a Christian or a Satanist, bear in mind, before you answer, that Christians of various persuasions vastly outnumber followers of Satan. This tendency of people to ignore the 'base rate' – how likely something is to occur in a random sample – when making a judgement given additional information is known as the base rate fallacy.

Of a similar type is the conjunction fallacy. In one study, carried out in 1983, subjects were told various things about a 31-year-old single woman named Linda, including that she majored in philosophy, had a strong concern for social justice, and took part in anti-nuclear demonstrations. They were then asked which was more likely, that: (1) Linda was a bank teller, or (2) Linda was a bank teller and a feminist. More than four-fifths of those questioned chose the second option, despite it being impossible for Linda to be a bank teller and a feminist without Linda also being a bank teller! Some people may have chosen option 2 because they thought it fitted better with other details they knew about Linda. Others may have avoided option 1 because they interpreted it as meaning 'bank teller but *not* also a feminist'. Before the men's final of Wimbledon in 1981, between Bjorn Borg and John McEnroe, people were asked which they considered more likely: that 'Borg will lose the first set' or that 'Borg will lose the first set but win the match.' The second option proved more popular even though, as in the case of Linda the bank teller, the first option includes the second as a subset. Seemingly, the majority made their choice based on which description they thought better represented what a top player like Borg would do in a match.

Given all the pitfalls we've talked about, scientists, who use statistics all the time in their work, need to be especially careful. Some form of statistical analysis is unavoidable whenever large amounts of data or numerous measurements are involved. Particle physics is a classic case where statistics plays a vital role in sifting through the outcomes of vast numbers of collisions in an accelerator looking for evidence of a new phenomenon or building block of nature.

An important measure of data gathered in an experiment is standard deviation, represented by the small Greek letter sigma, σ. Standard deviation is a measure of how widely the data are spread out around a central mean, or average. The smaller the standard deviation the more tightly the data are clustered about the mean. In particle physics, the sigma used is that of what's called a 'normal' distribution of data, which forms a bell-shaped curve. Two-thirds of the data in such a curve are contained within one standard deviation of the mean, 95 percent within two sigmas, and so on. Discovery of the long-sought-after Higgs boson came with the statement: 'We observe in our data clear signs of a new particle, at the level of 5 sigma, in the mass region around 126 GeV.' A spread of five standard deviations corresponds to a probability of 3×10^{-7}, or about 1 in 3.5 million. This isn't the probability that the Higgs boson does or doesn't exist but, rather, the probability that if it doesn't exist, the data collected would be at least as extreme as what was observed.

Researchers also have to be on guard against any kind of bias on the part of the experiment or experimenter. Say you're testing a new drug and need to decide if it actually works in human patients. The first step is to divide your sample of volunteers into two groups: a trial group and a control group. The trial group is given the actual drug, while the control group receives a placebo – an ineffective substance such as a sugar pill. None of those taking part knows which group they're in. Ideally, the study should be double blind, so that even the doctors administering the treatment don't know whether they're giving the real thing or the placebo. The main reason for this is the placebo effect: that merely *thinking* you received (or gave) the real treatment can have a positive effect.

Once all the data have been collected, they have to be analysed. The goal is to determine whether the trial drug actually works, or whether the results are simply what would be expected by chance alone. Here's where statistics takes an interesting approach. Instead of asking how likely it is that the drug works – what's called the alternative hypothesis – it's easier to figure out the likelihood that you'd get the same results if the drug *doesn't* work and is no better than a placebo. This option is referred to as the null hypothesis. If under the null hypothesis the results turn out to be unlikely, as measured by a predetermined test, the null hypothesis can be rejected and the claim made that the drug actually does what it says on the bottle. The strength of the claim depends on the statistical significance as measured by something called the confidence level. The most commonly used confidence level is 95 percent, which means that there's only a 5 percent, or 1 in 20, chance that the results could have been obtained under the null hypothesis. Those may sound like pretty good odds, but they still mean that for every twenty drugs that don't work, one may be reported by a study to be effective even though it really isn't.

After a significant result has been obtained and published, other researchers can then try to reproduce it, using the same method, to see whether it really was a coincidence. Only after much checking, and larger and more detailed trials, is a drug deemed safe to offer for general use.

Unfortunately, the pressure to publish encourages abuses

of statistics. When academics are seeking jobs, promotions, or funding for new projects, the volume of their published output – the number of papers in which their name appears as an author or co-author – is used as a performance indicator. What applies to individuals also applies to entire research groups: recognition and funding go hand in hand with results that appear in journals, especially prestigious journals that are held in high esteem around the world. This situation sets up potential conflicts because there's a tendency to tailor the content of papers to maximise their chances of publication, which can mean leaving out potentially interesting findings that challenge orthodoxy. After all, many of the referees used by journals to vet papers are authorities who may be inclined to look more favourably on research that falls in line with established science or their own understanding of the subject. There's pressure then to be cautious about putting forward anything that's too far out of the ordinary, even if it might involve a discovery of great interest. Worse is when researchers actually manipulate their data in order to fit some preconceived outcome by, for instance, not reporting measurements that fall out of line with the desired conclusion.

Another form of data manipulation involves testing a large number of different factors against each other. Assuming a 95 percent confidence level, 1 in 20 results on average will emerge as statistically significant even if, in fact, it eventually turns out that there's no correlation at all. If a researcher reports just this one connection that the statistics seem to suggest may be real, and at the same time fails to mention all the others that look statistically unfavourable, it gives the impression of significance where none necessarily exists. Such deception may ease the passage to publication but is no help at all to the scientific community.

Ultimately, science wins and makes progress because of peer review. When other, independent researchers reproduce a study, carrying out the same experiment using the same method, they'll determine, over time, whether the original result really is significant. If these follow-on studies find that the outcome is generally no better than the null hypothesis, then it'll be concluded eventually that the result isn't significant after all. The original researcher may have obtained what appears to be a significant result by coincidence (1 in 20 results are going to be significant by pure chance, even if the null hypothesis is correct), made some mistake in their methodology or data gathering, or have actively cheated by massaging the data in some way.

A common thread running through much of statistics is the idea of correlation. Two variables, say monthly rainfall and sales of umbrellas, are said to be positively correlated if, when one goes up, the other goes up, and vice versa; they're negatively correlated if the directions in which the variables change are in the opposite direction. The value of the correlation can range from +1 to -1. The end values represent 'perfect' or maximum correlation, where you can perfectly predict the value of one correlated variable when you know the value of the other.

It's easy to suppose that if two variables are correlated there must be a cause-and-effect relationship between them. But that isn't necessarily the case. For instance, the sales of ice cream in the UK are strongly correlated with the number of deaths by drowning. Does that mean eating ice cream affects your body in some way that makes it harder to stay afloat? Or perhaps that people resort to ice cream as a comfort food when hearing the sad news that someone has drowned? Not at all. Ice cream sales go up when it's hotter as does the number of people who go swimming. Despite the strong correlation between volume of ice cream sold and number of fatal drownings, it's the outside temperature that's the causal factor in both cases.

Often, though, it's not easy to decide if a correlation comes about because of a direct causal connection or not. Take the case of children who play violent video games. A policy statement by the American Psychological Association in 2015 said that research had shown a link: 'between violent video game use and both increases in aggressive behaviour... and decreases in prosocial behaviour, empathy, and moral engagement'. Some individual cases seem to support that view. An eighteen-year-old gunman who killed nine people in Munich in 2016 was reported to have been a fan of first-person shooter games. But correlation is no guarantee of causation.

A growing number of social scientists have called into question any causal link between watching violence and committing aggressive acts in real life. Most young people would probably agree with this survivor of the shootings at Stoneman Douglas High School, Florida, in 2018, in which seventeen people were killed: 'I grew up playing video games...first-person shooter games, and I would never, ever dream of taking the lives of any of my peers.' Even where there is a connection it may not be that watching triggers violence but rather that children who are predisposed to aggression also enjoy simulated violence on the screen.

There have even been suggestions that, at least in the short term, brutal video games help *reduce* violence in society. Psychologist Christopher Ferguson, at Stetson University, is among those who favour this idea. 'Basically, by keeping young males busy with things they like,' he said, 'you keep them off the streets and out of trouble.' A paper published in 2016 called 'Violent Video Games and Violent Crime' offered data to back up this claim. It concluded that in the weeks following the release of popular new video games that had a strong aggressive content, the level of general societal violence dipped.

Another well-known debate about correlation and causality sprang from research in the 1950s that pointed to a link between smoking cigarettes and lung cancer. The correlation between the two was clear. What couldn't easily be ruled out were alternatives to the theory that smoking was the direct cause of the cancer. It might be, for instance, that some pre-existing condition led to lung cancer in later life and also made the person more likely to smoke. Simply looking at the correlation couldn't rule out such possibilities. Also, there was no ethical way to test people over a long period of time by dividing them into groups: those who agreed to smoke, say, a pack of cigarettes every day for twenty years, and those who committed to abstain from smoking. Instead, the evidence of a causal link between smoking and cancer built up over time by various other means. It was noted, for instance, that smoking cigarettes was correlated with lung cancer, while smoking a pipe was correlated with lip cancer. It appeared that the development of cancer occurred in whichever body part came most into contact with tobacco smoke. This result, and similar ones, enabled over time a causal link to be established between cigarette smoking and lung cancer.

Triskaidekaphobics may feel that science has also vindicated the belief that unluckiness goes hand in hand with their most feared number – 13. In 1993, a group of researchers published results in the *British Medical Journal* on the number of accidents on the southern section of London's M25 motorway which involved injuries needing hospital attention. The study covered five months between 1990 and 1992 when the thirteenth happened to fall on a Friday. It compared the number of hospital-worthy accidents on the thirteenth with the number on the previous Friday for those months and found it to be 52 percent higher. If nothing else, the paper captured a little more attention in the popular press than do most scientific studies. But, as one of its authors, Robert Luben at Cambridge University's school of clinical medicine, pointed out, the whole piece was a very tongue-in-cheek affair. 'It was written', he said, 'for the Christmas edition of the *BMJ*, which usually carries fun or spoof articles.' Although the findings were real, they were never intended to be taken seriously. The numbers involved were so low that if there was any serious message at all it was to show how arbitrary statistics can be when small sample sizes are in play. Of course, that didn't stop the study from being taken at face value in some quarters and it continues to be cited, from time to time, as evidence that Friday the thirteenth is not a date to be trusted.

CHAPTER 9

Easier Said Than Done

> Some mathematics problems look simple, and you try
> them for a year or so, and then you try them for a hundred
> years, and it turns out that they're extremely hard to solve.
>
> – Andrew Wiles

HOW DID LIFE begin? Can we travel back in time? What came before the Big Bang? Why are we conscious? Some questions in science are so big and complicated that it may be centuries before we have the final answers to them – if we ever do. It's the same in maths. No one's expecting a solution any time soon to problems like the Riemann hypothesis (to do with the distribution of prime numbers) or P versus NP (whether every problem whose solution can be quickly verified can also be quickly solved). On the other hand, there are maths problems that look as if they should be easy to crack. They're certainly easy to understand and it seems that we ought to be able to resolve them pretty easily too. Yet despite the best efforts of mathematicians, in some cases for over a century or more, we've so far come up empty-handed.

In 1937, German mathematician Lothar Collatz made a surprising discovery. If you start with any whole number, divide it by 2 if it's even and triple it and add 1 if it's odd, and keep repeating this process over and over again, you'll eventually end up with the number 1. For instance, if you start with 20, and give it the Collatz treatment, you get the sequence 20, 10, 5, 16, 8, 4, 2, 1. Start with 17 and the sequence runs 17, 52, 26, 13, 40, 20, 10, 5, 16, 8, 4, 2, 1. The number of steps needed to get to 1 is called the total stopping time, and it can vary widely. For instance, if you start with the number 27 it takes 111 steps, during which time the highest value reached is 9,232 before the sequence reaches 1. But however many steps it takes, Collatz conjectured, 1 will *always* be the end point.

Needless to say, mathematicians have tried to find an exception, but so far they've drawn a blank. It's a simple

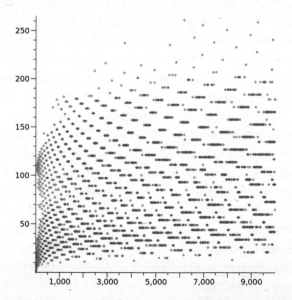

A plot of the numbers 1 to 9,999 (horizontal axis) against their Collatz total stopping time (vertical axis).

matter to program a computer to keep checking successively bigger numbers until it runs out of time or its owner out of patience. Mathematicians have applied this brute-force approach and shown (at the time of writing) that there are no violations of the hypothesis out to a value of 87×2^{60}. But this signifies nothing. The next biggest number that hasn't been checked might be the one that breaks the mould: that leads to a sequence that blows up to infinity or gets stuck in an endlessly repeating loop.

It wouldn't be the first conjecture that's been shown to fail at enormously large values. In 1919, Hungarian mathematician George Pólya suggested that more than half of the positive whole numbers less than any given number have an odd number of prime factors. But, in 1958, English mathematician Brian Haselgrove proved beyond doubt that Pólya's conjecture wasn't true, although he couldn't give any specific value that demonstrated this. The first explicit counterexample – 906,180,359 – was announced by American mathematician Sherman Lehman in 1960. Twenty years later, the smallest counterexample, just a tad less than Lehman's number at 906,150,257, was found by Japanese mathematician Minoru Tanaka, and, in fact, it's now known that Pólya's conjecture fails for the majority of numbers between 906,150,257 and 906,488,079.

Mertens' conjecture is named after Polish mathematician Franz Mertens but was first suggested by Dutchman Thomas Stieltjes in 1885. If it had turned out to be true, it would have rocked the mathematical world because it would have implied the truth of the mighty Riemann hypothesis. However, exactly a century after it was proposed, it was disproved. Shortly thereafter, it was shown that it must fail for some value between 10^{14} and $10^{1.39 \times 10^{64}}$, although no specific counterexample has yet been found. In any event, the experiences with Pólya's and Mertens' conjectures warn against any complacency that the Collatz conjecture must be true just because it's stood the test of time so far.

The beauty and mystery of what Collatz suggested is that anyone, even a child as young as eight or nine, can grasp what it's about. It's no exaggeration to say that it's the easiest-to-understand open problem in all of maths, having the appearance almost of a party trick. But the fact that it's remained unsolved since its discovery in Victorian times is a measure of its true depth. It relates not just to number theory but to issues of decidability, chaos, and the very foundations of mathematics and computation. Paul Erdős said of it: 'Mathematics is not yet ready for such problems.'

Several other easy-to-state but still open questions in maths have to do with prime numbers – numbers only divisible by themselves and 1. Twin primes, such as 3 and 5, 11 and 13, and 41 and 43, are primes that differ by only two. The twin prime conjecture is that there are infinitely many such pairings. It was first proposed in 1846 by French mathematician Alphonse de Polignac and so is sometimes referred to as the Polignac conjecture. No one made much progress on the theory of twin primes until Norwegian mathematician Viggo Brun came along in 1919. He showed that the sum of the reciprocals of the twin primes, $(1/3 + 1/5) + (1/5 + 1/7) + (1/11 + 1/13) + (1/17 + 1/19)\ldots$ gets ever closer to a fixed number as bigger and bigger pairs are included. Brun's constant, as this sum became known, was calculated in 1976, using twin primes up to 100 billion, to be approximately 1.90216054. In 1994, American mathematician Thomas Nicely was trying to refine the value further using a personal computer based on Intel's then-new Pentium chip, when he noticed he was getting some strange results. The problem was traced to a flaw in the chip, which Intel subsequently fixed. In 2010, Nicely extended the known accuracy of Brun's constant to $1.902160583209 \pm 0.000000000781$ based on all twin primes less than 20 thousand trillion (2×10^{16}).

An important step towards a proof of the twin primes conjecture came in 2003. American mathematician Daniel Goldston and Turkish mathematician Cem Yildirim showed that, if certain assumptions are made, there must be infinitely many twin primes that differ by no more than 16. Central among these assumptions is what's known as the Elliott–Halberstam conjecture – itself an open problem to do with the distribution of primes in arithmetic progressions (in which consecutive terms in a sequence of numbers differ by a constant amount). A mistake in their proof was corrected in 2005 with the help of Hungarian mathematician János Pintz. In 2013, American mathematician Yitang Zhang proved that, without making any assumptions at all, there's guaranteed to be an infinite number of prime pairs that differ by no more than 70 million. A year later this figure had been slashed to 246, and it could be cut further to 12 or 6, respectively, if the Elliott–Halberstam conjecture or a generalised form of it were assumed to be correct.

The Goldbach conjecture also has to do with primes and is very easy to grasp. Named after German mathematician Christian Goldbach, who corresponded on the subject in the eighteenth century with the great Leonhard Euler, it proposes that every even number greater than 2 can be expressed as the sum of two primes. For small numbers, it's obviously true: $4 = 2 + 2, 6 = 3 + 3, 8 = 3 + 5, 10 = 5 + 5$, and so on. The question is, does it continue to be true no matter how far down the number line we trek. We've already seen that some conjectures have held true all the way up to some enormously large value, then suddenly failed. Mathematicians aren't impressed by individual cases, no matter how many are put forward for the defence. They're not happy until they have a final, irrefutable proof, one way or the other. Goldbach's conjecture has been shown to hold for all integers up to 4 million trillion (4×10^{18}), and although mathematicians suspect that it's probably true for all numbers, that means nothing until a verified proof has been found.

Prime numbers are a tempting field of research for the unwary seeking quick mathematical fame. So many problems associated with them seem simple at first sight but turn out to be tremendously hard when you start to dig below the surface. Answer any of the following questions (complete with a proof acceptable to experts in the subject) and your celebrity as a number theorist is guaranteed: is there always a prime number between two consecutive squares? (Legendre's conjecture.) Are there infinitely many Fermat primes – primes that are 1 more than a power of 2? In fact, are there *any* such primes beyond 65,537? Are there infinitely many Mersenne primes – primes that are 1 *less* than a power of 2?

Many have ventured into prime territory, only to come away defeated by challenges that are far greater than they first appeared. Yet there's always hope of a major breakthrough, as British mathematician Andrew Wiles showed when he proved Fermat's last theorem in 1993, more than 350 years after it was first proposed.

Famously scribbled into the margin of a book by Pierre de Fermat in 1637, it was accompanied by the provocative comment: 'I have discovered a truly marvellous proof…which this margin is too narrow to contain.' Fermat's last theorem, known (more appropriately) in earlier times as Fermat's conjecture, states that no positive integers a, b, and c satisfy the equation $a^n + b^n = c^n$ for any integer values of n greater than 2. There are lots – in fact, an infinite number – of cases where the equation is satisfied for $n = 1$ or 2, for example, $3^2 + 4^2 = 5^2$, but you'd spend a long time trying to find values for a, b, and c that work if n is 3 or more. You'd also spend a long time trying to prove that there weren't any such values. When Andrew Wiles finally showed that Fermat had been right all along, it was by a circuitous and scarily difficult route involving a branch of maths called the modularity theorem for elliptic curves. There's no way – unless he'd had access to a time machine – that Fermat could have come up with a similar proof. Wiles had to pioneer new mathematical ground, and make breakthroughs that were much more important than vindicating Fermat, in order to get to his result. He made worldwide headlines, and is one of the few contemporary mathematicians many people can name, because he's the man who finally showed the last theorem to be right. For the far greater achievement of proving the Taniyama–Shimura conjecture for semi-stable elliptic curves, which would take an entire book to explain, he wouldn't even have got a mention in the popular press.

Very few mathematicians start out with the goal of becoming rich or famous. But for the solution of some maths problems, handsome financial rewards are on offer. One of these is the Beal conjecture – effectively a spin-off of Fermat's last theorem. Proposed by American banker and amateur mathematician Andrew Beal in 1993, it states that if $a^x + b^y = c^z$, where a, b, c, x, y, and z are whole numbers and x, y, and z are greater than 2, then a, b, and c have a common prime factor (a factor that's a prime number). There are probably easier ways of earning a million US dollars, but if you fancy having a go that's the prize currently on offer by Mr Beal for a successful, peer-reviewed proof (or disproof) of his proposition.

Solving Brocard's problem wouldn't earn you any money – except possibly that from a book contract – but you'd certainly be feted for it throughout the mathematical world. First posed by French mathematician and meteorologist Henri Brocard in 1876, and then, independently, by Srinivasa Ramanujan in 1913, it asks for integer solutions to the equation $n! + 1 = m^2$, where $n!$ is the factorial of n (for example, $5! = 5 \times 4 \times 3 \times 2 \times 1 = 120$). Only three pairs of numbers, called Brown numbers, are known that satisfy this equation, $(4, 5)$, $(5, 11)$, and $(7, 71)$, despite values of n up to a billion having been checked for further solutions. Try your hand at finding a fourth pair if you like, but it's strongly suspected, by Paul Erdős among others, that there isn't one.

If, after years of effort, you give up on the Brocard problem, how about turning your attention instead to this one: are there any odd perfect numbers? A perfect number is a whole number that's the sum of all the numbers that will divide into it (except itself). The smallest perfect number is $6 = 1 + 2 + 3$. The next one is $28 = 1 + 2 + 4 + 7 + 14$. They're not particularly common. The third and fourth perfect numbers are 496 and 8128. These first four were known to the Greeks as long ago as the fourth century BCE. But then there's a huge gap in both size and time before we find the next one – 33,550,336 – the earliest record of which is in a medieval German manuscript dating to 1456. At the time of writing, the biggest one known (the 51st) has more than 82 million digits and was discovered (by computer) in December 2018. Like all the other perfect numbers found so far, it's even. The possibility of an odd perfect number remains, though most number theorists would probably side with English mathematician James Sylvester who, in 1888, said:

> [A] prolonged meditation on the subject has satisfied
> me that the existence of any one such [odd perfect
> number] – its escape, so to say, from the complex web
> of conditions which hem it in on all sides – would
> be little short of a miracle.

Still in number theory, here's a surprising fact: we don't know if $\pi + e$ is rational or irrational. You might suppose it's bound to be irrational because both π (pi), the circumference of a circle divided by its diameter, and e, the base of natural logarithms, are irrational on their own. An irrational number can't be written as one whole number divided by another and in decimal form it goes on forever without repeating. It would seem only reasonable that if you added two such numbers together you'd get another one whose decimal expansion stretched out to infinity without any predictable pattern. But, in fact, it can be very hard to prove that a number is irrational. It's easier with some numbers than others. For instance, it's quite straightforward to prove that $\sqrt{2}$ is irrational. Proving the irrationality of e isn't that hard either, once you see how it's done, although it wasn't until 1737 that someone – Leonhard Euler – actually supplied a proof. But π is a different kettle of fish.

People have been aware of the properties of circles for several thousand years, so it's been known since ancient times that the circumference of a circle is slightly more than three times its diameter. Gradually, the value of π was figured out more and more accurately. Not until 1761, however, was pi's irrationality established, by Johann Lambert, based on an argument that involves continued fractions. If proving that individual numbers are irrational can be tricky, deciding on whether combinations of them are also irrational can be monumentally hard. This is especially so in the case of π and e, whose mathematical properties are so completely different. The problem's compounded by the fact that not only are π and e both irrational, they're also both transcendental. This means that neither is the solution of a polynomial equation (an equation such as $3x^3 - x^2 + 5x - 12 = 0$) with integer coefficients.

It seems likely that at some point we'll find out whether $\pi + e$ is irrational or not. Meanwhile, what we know for sure is that at least one of $\pi + e$ and $\pi \times e$ must be transcendental (and therefore irrational). It's also been shown, with the help of a Cray-2 supercomputer at NASA Ames Research Center in 1988, that neither $\pi + e$ nor $\pi \times e$ satisfy any polynomial of degree 8 or less with integer coefficients of average size less than a billion. Happily, we *do* know that e^π is transcendental, thanks to the work of Russian mathematician Aleksandr Gelfond. Before Gelfond produced the theorem that's now named after him, in 1934, only a few numbers, such as π and e, were known to be transcendental. But Gelfond's theorem opened the floodgates to establishing the transcendence of a whole class of numbers. It states simply that if a and b are algebraic numbers, and a doesn't equal either 0 or 1, and b is irrational, then a^b is a transcendental number.

If geometry is more to your taste, there's a whole slew of unsolved problems in this area of maths, too, waiting siren-like to attract your attention. One goes by the name of the 'happy ending problem' because two of the first mathematicians to work on it, Esther Klein and George Szekeres, met through their collaboration and eventually got married. As a young physics student in Budapest in the early 1930s, Esther was one of a small group, including Paul Erdős and future husband George, which met over new and interesting maths challenges. It was during one of their sessions that Esther asked the question that later became known as the happy ending problem. Although the problem itself doesn't yet have a happy ending, it's easy enough to describe.

Start by marking five random dots on a sheet of paper, with no three in a straight line. It turns out that you'll always be able to join four of them so as to form a convex quadrilateral – a four-sided shape in which all the corners are less than 180 degrees. To be sure of making a pentagon, you'd need nine dots; for a convex hexagon the minimum number of dots jumps to seventeen. But beyond that we're into unknown territory. No one knows the least number of random dots needed to ensure you could draw a convex shape with seven sides or more. The formula $m = 1 + 2^{n-2}$, where m is the minimum number of random dots required and n is the number of sides in the shape, works for the quadrilateral, pentagon, and hexagon. It's suspected it may continue to work for more complicated shapes, but no one's yet been able to prove it.

Here's another outstanding mystery, understandable with just pen and paper, to while away the hours. Draw a closed loop. It can be any shape you like, just so long as it's a simple loop that doesn't cross over itself and has no loose ends. Such a loop is called a Jordan curve. In 1911, German mathematician Otto Toeplitz put forward this proposal: every Jordan curve contains an inscribed square. Known as the Toeplitz conjecture, or inscribed square hypothesis, the suggestion is that inside a simple loop you can always draw a square so that all four corners touch the loop. It's easy to draw examples where this is true – a perfect circle being the most obvious case. But the question is, does the Toeplitz conjecture hold for every possible Jordan curve? You can amuse yourself all day drawing different loops and trying to find places where a square will exactly fit. But that proves nothing. What about the infinitely many other Jordan curves that you haven't had time to draw or test?

The Toeplitz conjecture has been shown to hold for a large class of Jordan curves, namely, those that are convex and smooth. The key point about a convex curve is that a straight line drawn between any two points on the curve will never cross over it. 'Smooth', in the mathematical sense, means that the slope of a line that just touches the curve (a tangent) never jumps abruptly from one point to the next; in plain language, the curve has no sharp corners. But Jordan curves

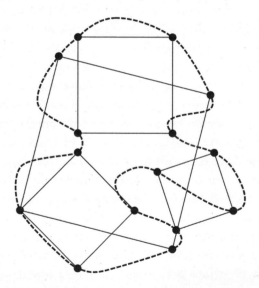

An example of the Toeplitz conjecture. The black dashed curve goes through all corners of several inscribed squares.

can be both concave and non-smooth, and this complicates the problem immensely. In fact, mathematicians have shown that the equivalent of the Toeplitz conjecture holds true for every type of Jordan curve in the case of a number of other shapes, including triangles and rectangles. But squares, it turns out, are trickier and so the hunt is still on for a proof that works for every possible curve, even those that have pointy bits or bow outwards.

Stepping up a dimension brings us to an unsolved problem that takes us back to our schooldays. At some point we all learn about Pythagoras's theorem: in a right-angled triangle, the square of the hypotenuse (the longest side) equals the sum of the squares of the other two sides. A Pythagorean triangle is a special case where all the sides have whole number lengths. Familiar examples are the (3, 4, 5) triangle, because $3^2 + 4^2 = 5^2$, and the (5, 12, 13) triangle. In three dimensions, Pythagoras's theorem works equally well but we need four numbers, to take account of the lengths in the x, y, and z directions plus that of the diagonal connecting the end points.

This brings us to the so-called perfect cuboid problem. Just as there are Pythagorean triangles where the lengths of all three sides are whole numbers, so there are some cuboids where the three edges (in the x, y, and z directions) and the spatial diagonal across the box – four lengths in all – have integer values. But each of the three faces of a cuboid also has a diagonal. The perfect cuboid problem asks: are there any boxes where all seven of these numbers, giving the lengths of the three edges plus the four diagonals, are whole numbers?

Mathematicians have come up with a few near misses. The shape known as an Euler brick is like a perfect cuboid but its spatial diagonal doesn't have to be an integer length. The easing of this restriction allows solutions to be found.

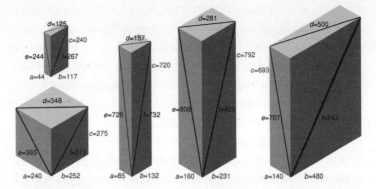

All five Euler bricks with dimensions under a thousand.

The smallest Euler brick was discovered, not by Euler himself, but by a contemporary of his, German mathematician Paul Halcke, in 1719. It has edges of length 44, 117, and 240, and face diagonals of length 125, 244, and 267. There are infinitely many Euler bricks and various methods are known for generating them. But the perfect cuboid, or perfect Euler brick, where the spatial diagonal must also be of integer length, has so far proved a stumbling block. Exhaustive searches have shown that the shortest edge of such a shape must be at least 5×10^{11} long. Various other conditions are known that a perfect cuboid must satisfy, such as that one side must have a length that's divisible by 5; one side, two face diagonals, and the spatial diagonal must be odd; and one edge or space diagonal must be divisible by 13. Although mathematicians haven't yet tracked down one of these beasts, they also haven't been able to show that none exists. The search continues...

Also unsolved but possibly of more practical value is the moving sofa problem. It really is called that and it's one with which anyone who's moved house is familiar: how to get a large, unwieldy shape around a tight corner. Mathematicians are a bit more precise in stating the problem than this and have also reduced it to two dimensions. The exact challenge, first put forward by Austrian-Canadian mathematician Leo Moser in 1966, is to figure out the largest 2D area that can negotiate the 90-degree bend in an L-shaped corridor that's 1 unit wide. This area, even in the halls of academe, is now fondly known as the sofa constant.

How to determine the value of this curious fundamental of the universe? If we start with something as simple as an armchair that's square when seen from above, measuring 1 by 1 (area = 1), it would be easy to steer round the corner. There wouldn't be much trouble either in negotiating the bend with a semi-circular seat of unit radius: just shove it to the end of the corridor, turn it through a right angle, and then pull it in the other direction. This would have an area of π/2, or about 1.571. But so far we've been lazy and chosen easy shapes. Time to get a bit more imaginative and mathematically creative.

In 1968, British mathematician John Hammersley came up with a funky new design – very 1960s – that pushed the record for the maximum area to π/2 + 2/π = 2.2074... The

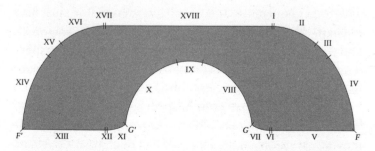

Gerver's sofa: the best solution found to date to the moving sofa problem.

Hammersley sofa, resembling, when seen from above, an old telephone handset, is made up of a rectangle with a semicircle cut out of it, and two quarter-circles stuck on the ends. It could actually be used as a sofa, as could a slight improvement on it found by Joseph Gerver in 1992. You'd be hard pressed to tell between the Hammersley and Gerver sofas at a casual glance. But Gerver's carefully crafted variant uses three straight sections and fifteen different curved sections that are nearly but not quite circular. It results in a very slight area increase to 2.21953167… No one's yet found a better solution and, though it remains unproven, it's quite possible that none exists.

Finding maxima or minima (a maximum area in the moving sofa case), subject to certain restraints, is the goal of optimisation problems. Not all such problems, however, have an answer in the traditional sense. A weird instance of this unusual situation is Kakeya's problem, posed by Japanese mathematician Soichi Kakeya in 1917.

Kakeya asked: what's the minimum area needed so that you could place a needle of length 1 inside the area and be able to rotate it through 180 degrees so that it gets back to its starting position (but pointing in the opposite direction)? You could spin the needle about its central point till it was upside down and the shape it travelled over would be a circle with a diameter of 1 unit and an area of about 0.785. But there are much smaller areas than this if the needle is pivoted in other ways. Kakeya explored various possibilities until he hit upon the deltoid – a shape that looks like an equilateral triangle but with sides that bulge inwards. A needle (or straight line of unit length) can pivot completely around inside this shape while only marking out an area of 0.393, half that of the circle. Kakeya believed that there was no shape more economical than the deltoid in which to spin his needle.

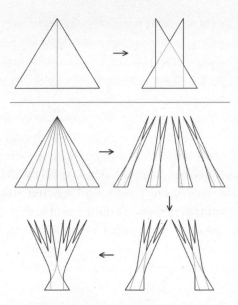

A way of approaching Kakeya's needle problem using a structure called a Perron tree. A triangle is divided into 2^n subtriangles, which are then recombined with partial overlaps. The cases where $n = 1$ and $n = 3$ are shown here.

But he was wrong. In 1919, Russian mathematician Abram Besicovich, in searching for a solution to Kakeya's problem that gave an even smaller area, made a startling discovery: there *is* no minimum area! Besicovich imagined taking an equilateral triangle, cutting it into a very large number of very thin strips, and then pushing these strips together so that they overlapped as much as possible. The result is a tree-like structure whose area, by cutting the triangle into thinner and thinner strips, can be made as small as we like. The 'branches' of the tree can also be connected to each other, again using an arbitrarily small area. A single tree like this allows Kakeya's needle to rotate through 60 degrees, so that gluing three of them together enables a 180-degree rotation. It's a very strange result, but true: the area required for a needle to make a half-spin can be as tiny as we like, providing it's not exactly zero.

There's a lesson then to be learned from this tale about easy-looking problems in maths: beware! Though they may look innocuous, simple-sounding questions to do with numbers or shapes can be like small openings to a vast cave. They can lead you into a previously unsuspected world of the mysterious, mystifying, and bizarre.

CHAPTER 10

Anything You Can Do...

> It is not enough to have a good mind. The main thing is to use it well.

> — René Descartes

EVERY YEAR, HUNDREDS of the brightest young mathematicians from around the planet compete in the International Mathematical Olympiad (IMO) – among the oldest and most prestigious scientific contests in the world. In 2018, one of the authors (Agnijo) managed to take joint first place in the IMO with a perfect score. Previous winners have included Maryam Mirzakhani, from Iran, who two decades later became the first woman to be awarded the Fields Medal – the equivalent in mathematics of a Nobel Prize – and Grigori Perelman, who later solved one of the most important open questions in maths: the Poincaré conjecture.

The first IMO took place in Romania in 1959 but competitiveness among mathematicians goes back a lot further. It became a serious business in the 1500s, in Renaissance Italy, when rival mathematicians would pose each other problems and try to solve them, while others would bet on the outcome. *Cartelli di matematica disfida* ('bills of mathematical challenge'), modelled on medieval jousts between knights, even featured juries, notaries, and witnesses. The winner might earn a new appointment or promotion at a university, although for most who competed the greatest prize was the fame and glory that success would bring, and, along with these, a bevy of new paying students. Not surprisingly, maths scholars tended to guard closely any discoveries they made because they could use them as weapons to defeat an opponent.

The most famous encounter of this kind took place in 1535 between Niccolò Fontana, nicknamed Tartaglia ('stutterer'), a self-taught but highly ambitious mathematician from Venice, and Antonio Fiore, a pupil of the eminent Bolognese professor Scipione del Ferro. Tartaglia's speech impediment stemmed from a brutal attack on him as a child, in which his jaw and palate had been cleaved with a sabre blow from a French soldier. The contest between Tartaglia and Fiore centred on finding solutions to cubic equations – equations in which the highest power of the unknown is three. It was already well known how to solve any quadratic equation, where the highest term was x squared, using a straightforward formula. But cubics proved a tougher nut to crack.

Del Ferro had secretly figured out how to solve equations of the form $x^3 + ax = b$ and passed this knowledge on, in strict confidence, to a handful of his friends and students, including Fiore. A few years after del Ferro's death, Tartaglia announced that he too had found a method for solving cubic equations. Determined to prove his superiority, Fiore publicly challenged Tartaglia to solve a series of cubic equations, confident of success because of the technique that had been passed on to him by his teacher. Unbeknown to Fiore, however, Tartaglia had gone one better than del Ferro and learned how to solve equations of the more difficult form $x^3 + ax^2 = b$, which included a quadratic term.

Under the rules of the challenge each contestant had to supply the other with thirty problems to be solved within forty days, the deadline being 22 February 1535. Unsurprisingly, all the problems that Tartaglia received were of the type $x^3 + ax = b$ (the only kind Fiore could solve), whereas Tartaglia, aware of Fiore's weakness, posed only problems that involved $x^3 + ax^2 = b$. The 'stutterer' breezed through every one of his questions in just a couple of hours, while his opponent was powerless, even after more than a month's effort, to attempt an answer to any of his.

Sadly for Tartaglia, this tale of intellectual one-upmanship didn't end well. A few years later he was persuaded to reveal his contest-winning formula to Milanese physician and math-ematician Gerolamo Cardano, after swearing him to secrecy. Typical of the period, Tartaglia conveyed his knowledge in an incredibly convoluted way – coded within the twenty-five verses of a poem that rhymed in the style of Dante Alighieri's *Divine Comedy*! Cardano, though, proved not to be a good secret-keeper. He passed on Tartaglia's trick to his protégé, Lodovico Ferrari, who, with Cardano's help, used it to solve quartics – equations involving an unknown raised to the *fourth* power. At first, Cardano and Ferrari were reluctant to publish their discovery, feeling duty-bound by the promise to Tartaglia. But, in the end, having become convinced that del Ferro had precedence over Tartaglia in pioneering solutions to cubics, Cardano relented and went public with general solutions for both cubics and quartics in his 1545 opus *Ars Magna* (The Great Art), acknowledging the contributions of both Tartaglia and del Ferro. Despite the credit given him, Tartaglia was furious about what he saw as a betrayal and, unfortunately, the whole thing descended into a slanging match between the two parties. It ended in another mathematical duel, this time between Tartaglia and Ferrari, which involved a total of six separate challenges over two years. In a final equation-solving slugfest in Milan, on 10 August 1548, before a large, partisan crowd massively favouring Ferrari, Tartaglia was humiliated – in part because the competition was oral and Tartaglia's speech issues made him an object of mockery.

On a personal level, of course, one can sympathise with anyone, like Tartaglia, who came off second best in these often-heated contests. On the other hand, there's no doubt that maths as a whole gained from the intense competitiveness and feverish mental activity involved when great minds went head to head. Ways of solving cubic and quartic equations were only the start. The formulae involved focused attention on the need to accept not only negative numbers as valid and respectable inhabitants of the mathematical universe but also an entirely new species, involving the square roots of negative numbers, which became known as complex numbers.

Today, there's still rivalry (mostly friendly!) between academics and between different research groups and academic institutions. Prizes, some of them quite lucrative, are on offer for those who, by common consent, are outstanding on the global stage or who can unravel one of the great unsolved riddles of maths, such as the mighty Riemann hypothesis. But actual competitions these days are generally restricted to students of school age.

The IMO itself technically allows anyone under the age of twenty to enter, although contestants mustn't have started university or any other tertiary education. Until 1966, only nations within the old Eastern bloc, such as Romania (the

Competitors at a mathematical Olympiad.

first host nation), Poland, Hungary, and the USSR took part. Thereafter, non-communist countries were invited to join the fun. 1967 saw the United Kingdom participate for the first time, along with Sweden, Italy, and France, and since then it's become increasingly globalised. Over 100 countries now take part every year, each sending a team of up to six contestants, plus a team leader and a deputy leader. The competition moves to a different country each year and spans about a week, including opening and closing ceremonies, lectures, trips out, and, of course, the nitty-gritty of the event – the problem-solving.

Six questions are given to participants, who work individually not as a team, three on each of two consecutive days in sessions that last four and a half hours. Each question is worth seven marks. Usually, any progress made without actually solving the problem is worth no more than three marks, whereas a solution with some fixable errors or one that's mostly complete may earn five or six marks, depending on the severity of the errors and omissions. Medals are awarded so that roughly the top twelfth receive a gold medal, the next sixth a silver, and the next quarter a bronze, while anyone who doesn't win a medal but gets full marks on one question achieves an Honourable Mention.

Questions 1 and 4, the first on each day, are regarded as the 'easy' ones but only because they're not so outrageously hard as the others! To most people they're still virtually impossible to understand let alone solve. Questions 2 and 5 are considered 'medium', and 3 and 6 'hard'. These hard problems are so seriously abstruse that only a handful of students each year end up fully solving both of them. To give an idea of just how tough they are, here's Question 3 from IMO 2017 held in Rio de Janeiro, which only two students fully solved and only seven in total scored any points on at all, making it, statistically, the toughest question ever asked:

A hunter and an invisible rabbit play a game in the Euclidean plane. The rabbit's starting point, A_0, and the hunter's starting point, B_0, are the same. After $n - 1$ rounds of the game, the rabbit is at point A_{n-1} and the hunter is at point B_{n-1}. In the nth round of the game, three things occur in order:

I The rabbit moves invisibly to a point A_n such that the distance between A_{n-1} and A_n is exactly 1.

II A tracking device reports a point P_n to the hunter. The only guarantee provided by the tracking device to the hunter is that the distance between P_n and A_n is 1.

III The hunter moves visibly to a point B_n such that the distance between B_{n-1} and B_n is exactly 1.

> Is it always possible, no matter how the rabbit moves, and no matter what points are reported by the tracking device, for the hunter to choose her moves so that after 10^9 rounds, she can ensure that the distance between her and the rabbit is at most 100?

IMO questions generally fall into the topic areas of algebra, geometry, combinatorics (dealing with combinations of objects), and number theory. In principle, they're solvable using maths, or extensions of maths, taught in high school. *In principle.* The fact is, even a first-rate high school maths education, by itself, would leave you pitifully ill-prepared to tackle any of the problems ever encountered in an Olympiad. Although you don't even need to know calculus to answer an IMO question, you do need an extraordinarily penetrating understanding of the connections between different theories and aspects of maths and a super-fast problem-solving ability that's beyond all but a tiny fraction of the population. Many competitors in the IMO, too, have a knowledge of maths, won through relentless study and practice, that goes way beyond what's taught in school or even at undergraduate level. This allows them to bring to bear tricks and shortcuts from left-field areas such as projective and complex geometry, functional equations, and advanced number theory. Most of those taking part will also have solved many thousands of problems, of increasing difficulty, during the course of their young lives, so they can spot when an IMO problem is similar to one they've already encountered and quickly form a solution strategy.

For the 1988 IMO, held in Canberra, a problem in number theory was proposed by West Germany that none of the six members of the Australian problem committee could solve, despite the fact that two of them, George and Esther Szekeres, were internationally renowned puzzle solvers and creators. Having baffled the committee, the problem was next sent to four of the top number theorists in Australia, who were asked to work on it for a maximum of six hours. None of these specialists could solve it either. Nevertheless, the problem committee passed the beast along to the jury of the 29th IMO flagged with a double asterisk – the mark of a problem so challenging that it might be too hard even for the Olympiad. After much deliberation, the jury allowed it in as the last question in the competition. It's also one of the shortest ever to appear in an IMO:

> Let a and b be positive integers and $k = (a^2 + b^2)/(1 + ab)$. Show that if k is an integer then k is a perfect square.

Remarkably, eleven students managed to solve it and gain a perfect score for this problem, including future Fields medallist Ngô Bao Châu from Vietnam, future Stanford maths professor Ravi Vakil from Canada, and future Mills College professor Zvezdelina Stankova from Bulgaria. (Remember, too, they had only four and a half hours to work on it plus a couple of other questions!) All used a technique that's become known as Vieta jumping after the sixteenth-century French mathematician and pioneer of modern algebra François Viète, whose Latinised name was Franciscus Vieta. One of the things Viète did was to discover formulae that relate the roots (or solutions) of a polynomial to its terms. For example, in the polynomial $x^2 + ax + b = 0$, the sum of the roots is $-a$ and the product is b. Those who managed to solve the infamous Question

6 realised that they could use Vieta's formulae in this new way, dubbed Vieta jumping.

Question 6 was especially tricky because it involves what's known as a Diophantine equation – an equation of which only integer, or whole number, solutions are allowed. These are often much harder to solve than ordinary polynomials where the roots can be any real (or even complex) numbers. To use Vieta jumping you start by setting up a quadratic equation and then looking for a 'minimal' solution, in other words, one that's as small as possible. You fix the smaller of *a* and *b* (which you can assume to be *b*, as *a* and *b* in the equation can be swapped around) and then find the two solutions of the quadratic. One solution has to be *a*, and Vieta's formulae can be used to find the properties of the other. It turns out, in the case of Question 6, that if the quotient isn't a perfect square, you end up with an infinite series of decreasing solutions, all in positive integers, which is impossible. On the other hand, if the quotient *is* a perfect square, you'll discover that at some point you hit 0 and the process stops. Just as in Renaissance times, when new maths could be forged in the intense heat of competition, IMO 1988 introduced a new weapon into the armoury of the number theorist.

Another of those who competed at the 1988 IMO was Terence Tao from Australia. Although Question 6 tripped him up, he still managed an overall points score of 34 out of a possible 42, enabling him to become the youngest IMO gold medallist at the age of only thirteen. In fact, Tao had already competed at two previous IMOs, taking a bronze medal as an eleven-year-old in 1986 (when he became the youngest-ever contestant) and a silver medal the following year. His exceptional ability was evident even earlier. As a nine-year-old, Tao began attending university lectures and scored above 700 points on the SAT math section – one of only two children so young ever to have done so. Today, he's regarded as one of the greatest of contemporary mathematicians, prolific in published output and extraordinarily capable across a wide range of topics.

A later-to-become-famous competitor in both the 1994 and 1995 IMOs, earning a combined total of 83 out of a possible 84 points, was the young Iranian mathematician Maryam Mirzakhani. After obtaining her BSc in maths at a university in Tehran, she studied for her doctorate at Harvard, where it's said she was 'distinguished by…determination and relentless questioning'. All of her class notes she took down in Persian. In time she became a professor at Princeton University and then Stanford before winning the 2014 Fields Medal – the first woman to do so – for 'outstanding contributions to the dynamics and geometry of Riemann surfaces and their moduli spaces'. In 2013, she was diagnosed with breast cancer, and she died, following its metastasis, in July 2017 at the age of forty.

AGNIJO'S IMO ADVENTURE

My own journey into the rarefied atmosphere of IMOs began long before I'd even heard of Maths Olympiads. UK primary school classes run from P1, for five-year-olds, to P7, after which children move on to secondary school and years S1 through S6. The teachers in the Scottish primary I attended spotted that I was good at maths and suggested, in my P5 year, when I was about nine, that I enter the Scottish Mathematical Challenge. The Primary category of the Challenge is intended

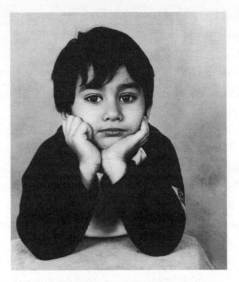

Agnijo as a youngster.

for P7 students and consists of three rounds, each with three questions, which are done at home and have a deadline of several weeks. Every year subsequently I took the Maths Challenge, quickly progressing through the various categories so that by my S2 year at high school I was competing in the Senior category intended for S5 and S6 students. I managed to solve every question each year, winning nine gold awards – a feat that hadn't been achieved before. In my parents' home there's now also a collection of nine Mathematical Challenge mugs, each in a different colour!

The Senior Mathematical Challenge is the first hurdle to overcome if you want to get into the UK IMO team. It's held in schools every November and consists of twenty-five progressively more difficult multiple-choice questions (marks being deducted for incorrect answers so that simply guessing doesn't pay). An example of an easy question is:

One of the following numbers is prime. Which one is it? A: $2017 - 2$, B: $2017 - 1$, C: 2017, D: $2017 + 1$, E: $2017 + 2$.

If you managed that one (answer: C), here's a harder one:

There is a set of straight lines in the plane such that each line intersects exactly 10 others. Which of the following could not be the number of lines in the set? A: 11, B: 12, C: 15, D: 16, E: 20.

The top thousand or so candidates in the Senior Mathematical Challenge (SMC) automatically qualify for the British Mathematical Olympiad Round 1 (BMO1), held in late November or early December and again sat in school, although candidates who don't automatically qualify can pay an entry fee.

The questions in BMO1 are in a very different format. There are six of them, all harder than anything posed at the SMC level and calling for that key ingredient present in all advanced maths and rarely dealt with in high school – rigorous proof. Each question carries ten marks and three and a half hours are allotted to work on them. Here's a sample BMO1 question if you'd like to have a go:

Consider sequences $a_1, a_2, a_3 \ldots$ of positive real numbers with $a_1 = 1$ and such that

$$a_{n+1} + a_n = (a_{n+1} - a_n)^2$$

for each positive integer n. How many possible values can a_{2017} take?

The top 130 candidates from BMO1 qualify for Round 2, although the cut-off marks are lower for students in earlier school years and, again, those that don't qualify can pay to take part. Sat in school in late January, BMO2 poses four questions, similar in style to those of BMO1 but harder – though sometimes the severity is sweetened with a little humour as in this gem from the 2018 paper:

> There are n places set for tea around a circular table, and every place has a small cake on a plate. Alice arrives first, sits at the table, and eats her cake (but it isn't very nice). Next the Mad Hatter arrives, and tells Alice that she will have a lonely tea party, and that she must keep on changing her seat, and each time she must eat the cake in front of her (if it has not yet been eaten). In fact, the Mad Hatter is very bossy, and tells Alice that, for $i = 1, 2 \ldots n - 1$, when she moves for the ith time, she must move a_i places, and he hands Alice the list of instructions $a_1, a_2 \ldots a_{n-1}$. Alice doesn't like the cakes, and she is free to choose, at every stage, whether to move clockwise or anticlockwise. For which values of n can the Mad Hatter force Alice to eat all the cakes?

The top twenty students from BMO1 (again, with lower cut-offs for younger contestants) attend a maths training camp held annually in the small town of Tátá, about forty miles from Budapest, along with the cream of young maths students from Hungary. The camp takes place over the New Year and lasts about a week. On most mornings there's a two-and-a-half-hour 'Individual Problem Solving' practice test, the rest of the day being filled with lectures by various speakers. On the final day, the British contingent takes a selection test, the outcome of which determines the half dozen students that form the team for the notoriously tough Romanian Master of Mathematics (RMM) competition. A few students whom the UK Mathematical Trust (the charitable body that runs Britain's national maths competitions) consider to be very strong candidates but who didn't qualify for the Hungary camp can take the selection test for the RMM at school, and, in fact, in 2018 half the team was chosen by this route. Anyone who makes the team for the RMM automatically qualifies for BMO2, even if they didn't get high enough marks on BMO1.

The top twenty or so contestants in BMO2 then move on to the next stage of the winnowing-down process – and you can no longer buy your way in! This next stage takes the form of another week-long training camp, held at Trinity College, Cambridge, again with lectures, mostly by former IMO contestants. There are two selection tests, on separate days, with the same format as the IMO – each involving three questions (harder than the BMO2 but easier than the IMO) and a four-and-a-half-hour period in which to solve them. Among the non-mathematical activities at the Trinity camp is that most traditional of Cambridge Uni pastimes – punting on the Cam.

At least five female students always attend the Trinity camp, four having already been chosen to form the UK team for the European Girls' Mathematical Olympiad plus one reserve. The tests given at the camp are used to select the teams for the IMO and the Balkan Mathematical Olympiad. Although obviously not a Balkan country, the UK takes part as a guest, and has a self-imposed rule that no one may participate more than once, so that a wider range of students has the chance to compete at Olympiad level. Those scoring in the top 8 to 10 at the Trinity camp, regardless of whether they've been in an Olympiad before, go forward as the IMO squad.

The final cut to decide the UK IMO team is made at a camp held, since 2015, at Tonbridge School, Kent, in late May. Four tests are given on consecutive mornings, each in IMO format and involving questions close to the IMO level of difficulty. The top six scorers form the team, with the rest of the squad making up the reserves.

I took part in the British Olympiad five times, starting in 2013 when I was twelve, and always progressed at least as far as BMO2. In 2016, I reached the Trinity camp and then again in 2018 – the year in which events happened in dizzying succession. In the space of a few months, the first book of our trilogy, *Weird Maths*, was published; I competed in three Olympiads, including the IMO itself; and I started my degree course at Cambridge.

The first of my trio of 2018 Olympiads was the Romanian Master of Mathematics, held in Bucharest in February. The format is the same as the IMO, but fewer teams take part (eighteen representing their countries, plus a B team from Romania and a team from Tudor Vianu High School, which hosts the competition). The questions at the RMM are more diverse than in other Olympiads and sometimes call for extra knowledge in areas such as calculus. They also have a reputation for being even harder than those at the IMO. I was happy to come away with a silver medal, ranking joint eleventh overall and joint first in the UK team, with 28 points out of 42. Funnily enough, before we boarded the plane to Bucharest, we walked into the airport branch of WH Smith's, and one member of our team spotted a copy of *Weird Maths*, asked me if I really was the co-author, and promptly bought it!

Agnijo at the 2018 International Mathematical Olympiad.

May saw me heading out with fellow teammates to the Balkan Mathematical Olympiad in Belgrade, Serbia. Because the competition clashed with the date of my Scottish Advanced Higher Physics exam, arrangements were made with the SQA (Scottish Qualifications Authority) for me to sit the exam in Belgrade at the same time as students back home – and the day before the competition. This time I won a bronze medal, having, like most of the UK team, found Question 1, a geometry problem, surprisingly hard. Balkan students tend to shine in geometry because their school curriculum puts a big emphasis on it, whereas my own strong points are combinatorics and number theory. Determined not to be caught out again, I ordered a copy of Evan Chen's *Euclidean Geometry in Mathematical Olympiads* and worked my way through its hundreds of problems involving the likes of cyclic quadrilaterals, mixtilinear incircles, and projective transformations.

The time of the IMO was now fast approaching. Along with the rest of the team I flew out to Budapest for one last, pre-IMO training camp, held jointly, as it is every year, with the Australian team. On each of the five days we were given a practice paper in the IMO format of three questions and four and a half hours to solve them. On the third day, the British team set a paper for the Australian team and marked it, and vice versa. The fifth and final paper was for the 'Mathematics Ashes', a contest between the two teams that began in 2008. It mirrors the slightly more famous Ashes contests between England and Australia at cricket in which an urn, or replica of it, purportedly containing the burnt remains of a bail from a long-ago contest, changes hands depending on who's won the latest Test series. The Mathematics Ashes urn contains a burnt answer script and has been won every year by the UK team except for the year of its inception, 2008, and 2018, when we lost by one point despite two of us (including me) getting a perfect score.

A couple of days after returning home, the UK team was on its travels again – this time to the IMO itself. Fifty-nine years after it started there, the IMO had returned, for the sixth time, to Romania and to the city of Cluj-Napoca, the country's fourth most populous. In total, 594 contestants assembled from 107 countries. At the opening ceremony, on Sunday 8 July, speeches were given by the President and Vice Prime Minister of Romania, the Mayor of Cluj-Napoca, and the President of the IMO Committee, Geoff Smith (who also happened to be the UK Team Leader), followed by a parade of all the teams.

The competition proper came on the next two days, with an hour's delay at the start for security reasons. On the Tuesday, the Finnish team decided to take advantage of this wait to have a lie-down in the middle of the exam hall. Other teams, interpreting this as a protest against the long delay started lying down as well, until there was a line of people that spanned the entire room. Next, we got up and started marching around the hall before some members of the UK team picked up the IMO flag and waved it as they went. The other teams took their own flags and marched as well. Eventually, we were asked to take our seats for the test, at which point the invigilator thanked us for the unofficial show.

To give a flavour, here are a couple of questions from the first day's set:

> QUESTION 1: Let Γ [gamma] be the circumcircle of acute-angled triangle ABC. Points D and E lie on segments AB and AC, respectively, such that $AD = AE$. The perpendicular bisectors of BD and CE intersect the minor arcs AB and AC at points F and G, respectively. Prove that the lines DE and FG are parallel (or the same line).

> QUESTION 3: An anti-Pascal triangle is an equilateral triangular array of numbers such that, except for the numbers in the bottom row, each number is the absolute value of the difference of the two numbers immediately below it. For example, the following array is an anti-Pascal triangle with four rows which contains every integer from 1 to 10.

$$
\begin{array}{c}
4 \\
2\ 6 \\
5\ 7\ 1 \\
8\ 3\ 10\ 9
\end{array}
$$

> Does there exist an anti-Pascal triangle with 2018 rows which contains every integer from 1 to $1 + 2 + \ldots + 2018$?

These problems, and others like them, can be tackled in many different ways, but the typical approach in school maths of simply using some formula and working through the calculations won't get you very far. At IMO level the emphasis is on coming up with watertight proofs, but these can only be obtained following some insight into the solution method. That insight, in turn, depends on years of practice plus the injection of some creativity at the personal level.

The contests are both intensive and fun for everyone involved – meetings of like-minded maths fans who get a lot out of being in each other's company and challenging each other's wits. The bonding and companionship is helped by excursions that follow in the days after the contest itself. On the first of these we went to the nearby city of Alba Iulia, visiting its two cathedrals, Eastern Orthodox and Catholic, and Salina Turda (the Turda salt mine). A hundred metres underground we played a round of Mao, a card game popular at Olympiads and training camps, in which the aim is to figure out the rules, except, in the salt mine, there was an added twist, literally – we played it on a roundabout.

While the teams were doing the fun stuff after their hard work, the leaders and deputy leaders took part in 'coordination', where they presented their team's solutions and how many marks they thought they were worth. These recommendations were then passed on to the coordinators, who decided the final scores.

With the final day came the closing ceremony and, again, a round of speeches by notable figures followed by a list of all Honourable Mentions – those competitors who, despite not having won a medal, had scored full marks on at least one of the questions. Then the medal winners were invited onto the stage, in groups of twelve, starting with the bronze medallists (in ascending order of score), followed by silver and gold. Finally, it was my turn to go onstage along with the other perfect scorer, James Lin of the USA team. I'd managed to achieve full marks on all six questions – the first time since 1994 that anyone on the UK team had scored a perfect 42. The Romanian team then handed over the IMO flag to the UK team in recognition of the fact that the United Kingdom would host the 2019 IMO, in Bath.

Although success at IMO 2018 was an amazing (and unexpected!) climax to my schooldays, it was only a first step into the strange and astonishing world of advanced maths. For me, now studying at Trinity College, Cambridge, and aiming at a career in mathematical research, the adventure has just begun.

CHAPTER 11

Logic: Formal, Fallacious, and Fuzzy

> Contrariwise, if it was so, it might be; and if it were so,
> it would be; but as it isn't, it ain't. That's logic.
>
> – Lewis Carroll

LOGIC IS...HARD TO define, no matter how logical you try to be about it. The word comes from the Greek *logos*, which the ancient Greeks themselves used in an impressive and confusing number of different ways. It might mean 'word', 'reason', 'speech', 'opinion', 'plea', 'plan', or half a dozen other things, depending on your philosophical point of view.

Early on, logic was considered to be one of the pillars of rhetoric – the art of convincing others by skilful use of language – along with *ethos* and *pathos*. To some philosophers, the most important aspect of *logos* was its effectiveness as a tool in winning arguments. Others said it had nothing to do with a speaker's powers of persuasion, but was instead nothing less than the inherent order and meaning of the universe. In the case of Christianity, this cosmic code was personified in the form of a supposedly divine individual: literally, the word of God as conveyed by Jesus Christ.

Our modern understanding of logic stems from ideas that were set down by Aristotle in six works known collectively as the *Organon*. This was the first comprehensive work in the Western world on what's become known as formal logic – a system for determining if the conclusion of a line of reasoning is valid or not. Aristotle started from basic elements or subjects of discourse such as 'men', 'mortal', and 'Socrates', which in themselves are neither true nor false: they just are. These raw elements, he explained, can be built into propositions, such as 'All men are mortal' or 'Socrates is a man.' Put three propositions together – two premises and a conclusion – and you get what's called a syllogism. A classic one of Aristotle's is:

> All men are mortal.
> Socrates is a man.
> Therefore, Socrates is mortal.

A syllogism on its own is just a simple argument, in the Aristotelian sense. More sophisticated arguments can be pieced together from the conclusions of different syllogisms. But a syllogism and, by extension, an argument, isn't necessarily valid. The one just mentioned is because it's true that *if* all men are mortal and *if* Socrates is a man, then it must be the case that Socrates is mortal. But think about this syllogism:

> All men are mortal.
> All cows are mortal.
> Therefore, all cows are men.

It's invalid even though there's nothing wrong with the first two premises. Whether a syllogism is valid or invalid depends entirely on the structure of the argument and not on whether all or any of the premises are actually true. For instance, the following syllogism, while logically valid, is also utter nonsense:

> All men are cows.
> All cows are dolphins.
> Therefore, all men are dolphins.

It's nonsense because, although there's nothing wrong with the logic, neither of the two premises is actually true! This is where the concept of 'soundness' comes in. A syllogism or argument is said to be sound if it's valid *and* all of the premises that underpin it are true. Logic on its own isn't enough to tell us whether an argument's sound – we also have to know about the premises. It's only when we're happy to accept that the premises are true that we can be confident that the logic based on them will guide us to a valid conclusion.

Aristotle's *Organon* goes into a tremendous amount of detail, especially regarding his theory of syllogisms. It talks about various types of proposition that can be used in arguments, not just ones that conclude 'All A are B' but also others that state 'Some A are B' or 'Some A are not B'. Aristotle was the first to deal, in a systematic way, with concepts such as the excluded middle – the idea that no matter what the proposition, either it must be true or its negation must be true. His work had a tremendous influence on future Western thought.

But important developments in logic weren't confined to the West. In the seventh century BCE, concurrent with the rise of philosophy in ancient Greece, the doctrine of *anviksiki* took root in India. The Sanskrit term 'anviksiki' roughly translates as the 'science of inquiry', and in its earliest form, around 650 BCE, embraced both the science of the soul and the theory of reason. The development from it of Indian logic is credited to Medhatithi Gautama, who wrote an early text on the subject in about 550 BCE. There are overlaps between Aristotelian and Indian logic but the latter is much more concerned with philosophical ideas of the mind and the senses. Whereas Aristotle, and Western logic in general, takes a purely materialistic and objective approach, Indian logic assumes from the outset that subject and object are inseparable and intertwined. This leads to its conclusion that objects have qualities that don't have an existence of their own. Indian logic, too, is much more analytical of concepts like perception. It takes a keen interest in the process of perception, questioning the independent existence of the object being perceived, the involvement of the medium (such as light in the case of seeing), and the nature of the sense organs, mind, and self. Not all perceptions, it concludes, are equally valid or necessarily valid at all, so that any presumed logic flowing from them may be equally suspect.

A third strand of logic evolved in China, beginning with Confucius between the sixth and fifth centuries BCE. But Confucianism was massively suppressed during the first Chinese imperial dynasty – the Qin dynasty – towards the end of the third century BCE, and afterwards the ideas about logic adopted in China were mainly those imported from neighbouring India. Certainly, in terms of their impact on mathematics, it's the systems of logic that grew up in India and the West that are by far the most significant.

Greeks other than Aristotle played a part in logic's development. Members of the Stoic school of philosophy, most notably Chrysippus in the third century BCE, were behind the growth of what's known as propositional logic, a distinct branch from the predicate logic founded by Aristotle. As the name suggests, the key elements of propositional logic are propositions – statements about the real world that are either true or false. From a set of initial propositions, new ones can be built using logical connectives. For instance, starting with the propositions 'there's a cat in Nancy's house' and 'Nancy is Freddie's girlfriend' you can make the proposition 'there's a cat in Freddie's girlfriend's house.' If propositions are like atoms, then predicate logic works more at the subatomic level, enabling the analysis of the internal structure of propositions.

As mentioned earlier, just because an argument makes logical sense doesn't guarantee that it's true. All sorts of fallacies come about even when the chain of logic looks fine. For instance, using a bit of basic algebra and what looks like a logical sequence of steps it's easy to prove that 2 equals 1.

Start by putting $a = b$.
Multiply both sides by a: $a^2 = ab$
Add a^2 to both sides: $a^2 + a^2 = a^2 + ab$
So $2a^2 = a^2 + ab$
Take $2ab$ from both sides: $2a^2 - 2ab = a^2 + ab - 2ab$
So $2a^2 - 2ab = a^2 - ab$
Or $2(a^2 - ab) = 1(a^2 - ab)$

Finally, divide both sides by $(a^2 - ab)$, and we arrive at the startling conclusion

$2 = 1$.

Things look bad for the whole edifice of maths at this stage, until we realise what was done in that last step. Earlier, we wrote $a^2 = ab$, from which it follows that $a^2 - ab = 0$. A long time ago, mathematicians realised that they couldn't allow division by zero because otherwise they ended up with all kinds of nonsense, such as being able to show that any number is equal to any other number. Although the logic of our $2 = 1$ argument looks fine, it leads to a fallacy because dividing by zero is undefined.

The principles of Greek logic, particularly those set out in the *Organon*, formed the basis for further developments by Islamic and medieval European philosophers, such as William of Ockham and Jean Buridan. A long fallow period followed, stretching from the fourteenth century to the early part of the nineteenth. But then interest in logic revived as it began to be seen as fundamental, not only to reasoning as a whole, but also to the foundations of science and maths in particular. Central among those involved in this renaissance were British mathematicians George Boole, Augustus De Morgan, and Charles Babbage.

It was around the end of the eighteenth century and the turn of the nineteenth that some British scholars also began to take an interest in Indian philosophy. They awakened, in the West, an understanding of the sophistication of the systems of inference and analysis that had sprung up more than two thousand years earlier on the subcontinent. Aristotelian logic had dominated for so long in Occidental thought that it came as a revelation to many that an entirely different, and in some ways more subtle, scheme of logic had sprung up in the East.

The study of logic, both Eastern and Western, began to spread beyond the bounds of philosophy and into mathematics. The two had always been close in nature, both firmly dependent on meticulous, step-by-step reasoning. As De Morgan wrote in 1860: 'The two races which have founded mathematics, those of the Sanscrit [*sic*] and Greek languages, have been the two which have independently formed systems of logic.' But throughout history, up to this point, maths and logic had developed more or less separately along parallel tracks. It was only towards the middle of the nineteenth century that the idea of *quantifying* elements of logic arose and with it the birth of mathematical logic.

Augustus De Morgan was prodigiously bright and knowledgeable, even as a child. He knew Latin, Greek, and some Hebrew by the age of fourteen and entered Trinity College, Cambridge, to study maths as a sixteen-year-old. The huge amount of time he spent reading on every subject under the sun got somewhat in the way of his studies and he came 'only' fourth in the mathematical tripos. Also, his refusal to sit the theological tests then required in order to take a master's degree meant he was barred from joining the faculty at Cambridge. But he did go on to teach at the newly established London University, where he gained a reputation for being a brilliant lecturer.

In his twenties, De Morgan started to take a serious interest in bringing logic and maths together. He wasn't alone in this. During his time at Cambridge one of his tutors had been the polymath William Whewell (inventor of the word 'scientist'), who argued that science needed to embrace the principles of logic more in its methodology. De Morgan's focus was on quantifying logic, starting with the use of symbols to represent the various parts of syllogisms, so that logical propositions and conclusions could be transformed into what looked like expressions in algebra. He struck up a busy correspondence with others who were exploring similar ideas at the time, including George Boole, the Scottish philosopher Sir William Hamilton, and the Irish mathematician, with (confusingly) almost the same name, Sir William Rowan Hamilton.

Both Boole and De Morgan had steeped themselves in

Aristotelian logic and the (to Western eyes) exotic xeno-logic of the East. De Morgan wrote openly about the contributions of Indian logicians, while Boole's knowledge of the subject was affirmed by his wife, Mary Everest Boole, in an essay titled 'Indian Thought and Western Science in the Nineteenth Century'. The linking factor, claimed Mary, was her uncle, George Everest, after whom the world's highest mountain is named, and who, having lived for a long time in India, brought back to England many ideas, including philosophical ones, that he'd picked up from the region. In her essay, she wrote: 'Think what must have been the effect of the intense Hinduising of three such men as Babbage, De Morgan, and George Boole on the mathematical atmosphere of 1830–65.' One effect may have been to encourage these pioneers of a more modern wave of thinking to question the shortcomings of classic propositional logic and move logic in a more mathematical direction.

George Boole was largely self-taught and spent most of his career in Ireland as the first professor of mathematics at Queen's College, Cork. His death, at the age of only forty-nine, followed an unfortunate sequence of events that began with a three-mile walk in a downpour from his house to the university and a subsequent lecture conducted in his saturated clothing. Subsequently, he developed pneumonia, which possibly he may have survived had it not been for the doubtless well-intentioned ministrations of his wife. Mary, while a first-class thinker in other respects, as her mathematical writings attest, was under the illusion that cures should match causes and therefore wrapped George in wet blankets. After a couple of weeks of this treatment, his immune system gave up the unequal struggle and he expired of pleural effusion on 8 December 1864.

A decade earlier, Boole had published his ground-breaking book *An Investigation of the Laws of Thought*, in which he wrote:

> There is not only a close analogy between the operations of the mind in general reasoning and its operations in the particular science of Algebra, but there is to a considerable extent an exact agreement in the laws by which the two classes of operations are conducted.

Boole devised what's become known as Boolean algebra, or Boolean logic, a mathematical system for working with variables that can have only two possible values: true or false. Whereas the four main operations in ordinary arithmetic are add, subtract, multiply, and divide, the primary operators of Boolean algebra are AND, OR, and NOT. The actions of these operators are shown in so-called truth tables, which describe the output for every combination of inputs.

A	B	A AND B	A OR B	NOT A
FALSE	FALSE	FALSE	FALSE	TRUE
FALSE	TRUE	FALSE	TRUE	TRUE
TRUE	FALSE	FALSE	TRUE	FALSE
TRUE	TRUE	TRUE	TRUE	FALSE

By linking together the operators AND, OR, and NOT in various ways, more complex logical statements can be assembled, such as ((NOT A) OR C) AND NOT (A AND (NOT B)). The three basic operators can also be used to construct other operators. Important among these are NAND (not AND), NOR (not OR), and XOR (exclusive OR). For variables A and B, A NAND B = NOT (A AND B), which is true whenever at least one of A and B is false; A NOR B = NOT (A OR B), which is true whenever both A and B are false; and A XOR B = (A OR B) AND (A NAND B), which is true whenever exactly one of A and B is true.

In time, Boolean algebra became one of the cornerstones of modern computing. Electronic computers, at the most basic level, consist of large numbers of switches, each of which can be either on or off. These two states can represent the 'true' or 'false' values in Boolean algebra, or the digits '1' and '0' in binary arithmetic. Many of the circuits inside computers consist of switches that are arranged to make logic gates, which correspond to the various operators in Boolean algebra. Each type of gate gives an electrical signal as an output depending on which signals it receives from its inputs: a 'true' or '1' corresponding to a high signal and a 'false' or '0' corresponding to a low signal. In today's computers, the numbers of logic gates involved and the complexity of their arrangements are tremendous – the latest chips, as small as a fingernail, may hold a billion gates and have a transistor (switch) density of 100 million per square millimetre.

George Boole couldn't have foreseen where his innovative new form of logic would lead. But he did witness some early developments in computers. In 1862, Boole met Charles Babbage at the Great London Exposition in South Kensington. On display were parts of Babbage's Analytical Engine, which was intended to be a successor to his Difference Engine, a mechanical calculator designed to work out automatically, with the turn of a crank, the values of functions, such as sines, cosines, tangents, and logarithms. Babbage had managed to complete a small working model of the Difference

The London Science Museum's Difference Engine, the first one actually built from Babbage's design.

Engine and demonstrated it to Lady Byron and her daughter Ada Lovelace in 1833. In her diary, Lady Byron wrote: 'We both went to see the thinking machine (for so it seems) last Monday. It raised several Nos. to the 2nd and 3rd powers, and extracted the root of a Quadratic equation.'

The Analytical Engine would have been a huge leap forward even from the Difference Engine, which was itself far more advanced than any previous calculating device. Babbage's concept was for a truly general-purpose mechanical calculator, whose instructions and data would be fed in via loops of punched cards like those used in the Jacquard loom. It was never built for the same reason that a full-scale Difference Engine was never completed – although the design was sound, the practical requirements for its construction, in terms of engineering tolerances, were beyond the technology of the time. Nevertheless, Boole came away from his meeting with Babbage impressed by what he'd seen and heard. In a letter to Babbage, dated 15 October 1862, he wrote:

> My dear Sir, It is a source of regret to me that I was quite unable to avail myself of your kind invitation to call upon you on my return from Cambridge to London... Meanwhile, I shall endeavour to acquaint myself with Menabrea's paper and the principle of the Jacquard loom.

Although Boole couldn't have dreamed that in little over a century his work on logic would form the basis for the digital world in which we now live, at least he was there at the earliest dawn of the computer age.

After Boole, logic was taken even further in a mathematical direction by German philosopher and logician Gottlob Frege. His goal was nothing less than to show that arithmetic and logic are identical and that the concept of number can be defined by purely logical means. Others joined the effort to try to prove that every branch of maths, from geometry to set theory, would arise as a matter of course from a collection of axioms, or self-evident statements, laid down at the outset. Frege's early work on what's called naïve set theory was derailed by a paradox found by English philosopher Bertrand Russell in 1901 (as we explain in one of the chapters of *Weird Maths*). But German logician and mathematician Ernst Zermelo found a way of resolving the paradox, and together with German-born Israeli mathematician Abraham Fraenkel, built a system of axioms – Zermelo–Fraenkel set theory – which, along with the axiom of choice, forms the most commonly accepted logical foundation of mathematics.

At the dawn of the twentieth century, there was optimism about being able to ground all existing maths in a set of axioms that could be proved to be consistent. The effort to do this was crystallised in Hilbert's programme, named after German mathematician David Hilbert. However, Hilbert's dream was dealt a fatal blow, in the eyes of most (but not all) mathematicians, in the 1930s by Austrian-American logician Kurt Gödel with his incompleteness theorems (which, again, we explore in *Weird Maths*).

A lot of the logic used in today's science, maths, communication, and computation depends on principles originally laid down by Aristotle and developed much later into quantitative form by the likes of De Morgan and Boole. But there are many situations where two-state classical logic, based on true or false, isn't appropriate. Suppose a hot object starts to cool down – at what point does it stop being hot and become cold? Clearly, there's no sudden switchover: no specific temperature at which it abruptly becomes false to say that's it hot. To deal with cases like this we need a different approach.

Aristotle himself was aware of some of the limitations of two-value logic and talked about them in *De Interpretatione*, the second book of his *Organon*. He realised that his law of the excluded middle was on shaky ground when, for instance, it was applied to future events because these are contingent – not yet either true or false. To illustrate the problem he came up with a thought experiment. Suppose there isn't going to be a sea battle tomorrow (sea battles being more of an issue in the fourth century BCE than they are today!). Then this must also have been true in all previous times because anything that's true about a future event must also have been true about that event in the past. But the necessity of all past truths up to the statement that 'A sea battle will not be fought tomorrow' implies that the opposite statement, that one *will* be fought, is necessarily false. Therefore, it isn't possible that a battle will be fought. Aristotle put it this way:

> For a man may predict an event ten thousand years beforehand, and another may predict the reverse; that which was truly predicted at the moment in the past will of necessity take place in the fullness of time.

As far as the case of the sea battle goes, it's impossible for both alternatives to be possible at the same time: either there *will* be a battle or there won't. Aristotle's solution was to argue that today, neither proposition is true or false; but if one is true, then the other becomes false. It's impossible to say today if the proposition is correct: we must wait for the contingent realisation (or not) of the battle; logic then realises itself afterwards.

Although he recognised that there were situations in which two-valued logic struggled, Aristotle didn't make any suggestions about going beyond it. In fact it wasn't until around 1917 that the first non-classical, many-valued logic system was devised, by Polish logician and philosopher Jan Łukasiewicz. To deal with Aristotle's paradox of the sea battle, Łukasiewicz introduced a third truth value, 'possible'. In later years, he and other logicians, including fellow Pole Alfred Tarski and American mathematicians Emil Post and Stephen Kleene, went further in exploring multi-valued logic having n different truth values, where n is greater than 2. In 1932, Łukasiewicz and his student Tarski became the first to entertain the notion of infinitely many-valued logic, from which in time, the idea of 'fuzzy logic' emerged.

The distinguishing feature of fuzzy logic is that a statement doesn't have to be entirely true or false, or, in numerical form, 1 or 0. Instead, it can take any value in between, so that, for example, a statement with a truth value of 0.8 would be mostly true. Fuzzy logic is often better at describing how humans use certain terms. In the case of temperature, we talk about things becoming gradually warmer or cooler without there being sharp cut-off points. In fuzzy logic, if the temperature of an object rises, this would equate to an increase in the truth value of 'It is hot' and a corresponding decrease in the truth value of 'It is cold.'

Fuzzy logic can also prevent the so-called sorites paradox, or heap paradox, which was first discussed by Eubulides of Miletus ('sorites' is the Greek word for heap). The original version of it supposes that there's a heap of sand from which one grain is removed: obviously, it'll still be a heap. But keep repeating this process, removing grains one at a time, and, eventually, there'll be just a single grain left – which is clearly not a heap. In Aristotelian two-state logic the only way to prevent this naming problem from arising is, at each step, to define whether what's left is a heap or not, so that there's an arbitrary point of transition at which a mound of sand grains ceases to qualify as a heap. But then we end up with the absurd situation of two piles of sand, differing in size by a single sand grain, one of which qualifies as a heap but the other doesn't. Fuzzy logic enables a way around this ridiculously artificial jump from one state to another. It formalises the fact that you can instead say that removing one grain of sand from a heap makes it very, very slightly less than a heap. The truth value of the statement 'This is a heap' is then allowed to decrease gradually as you remove grains of sand, so that a small pile of sand is identified as being less of a heap than a larger pile.

The sorites paradox pops up in countless other guises – for example, concerning the subject of wealth. A billionaire is obviously rich. Removing one penny from a rich person's wealth doesn't stop them from being rich. But insist on retaining that label, as penny by penny disappears, and in time you'd be forced to conclude that someone who's completely bankrupt is still rich! Applying fuzzy logic allows the more reasonable position that losing even a single penny reduces the truth value of the statement 'This person is rich' by a tiny but non-zero amount. Fuzzy logic also allows us to interpret words like 'quite' or 'slightly' in terms of a range of truth values.

The term 'fuzzy logic', and the maths behind it, was put forward in the 1960s by Azerbaijani-born mathematician and computer scientist Lotfi Zadeh of the University of California at Berkeley. At the time, Zadeh was working on the problem of how to enable computers to understand natural language. Like most things in life, natural language often doesn't lend itself to absolute terms like 'true' and 'false' or the kind of logic that imposes a choice between such extremes. With fuzzy logic, true and false, or 1 and 0, are included at the opposite ends of a truth scale that provides for an infinitely fine gradation of truth in between. For example, in gauging a person's height it doesn't insist on a choice between 'tall' and 'short' but allows for statements such as '0.43 of tallness'.

Making incremental and provisional assessments is part of the modus operandi of human thought. Researchers therefore see fuzzy logic as a way of incorporating such nuanced, human-like capabilities into artificial systems. This has led to software being developed that, when faced with an unfamiliar task, goes about finding a solution broadly along the same cognitive lines as we do.

Another feature of our brains is that they automatically, and at a subconscious level, group together data and form various provisional results, which are further combined to yield broader and more strongly weighted conclusions. When these conclusions reach a certain threshold, 'we' may become aware of them and, as a result, make a conscious decision to act in some way. Alternatively, the aggregated output may automatically trigger some other external behaviour such as an involuntary motor reaction. Efforts to mimic this process of consensus building – progressively combining and weighting the output from many data processing elements – have led to the construction of artificial neural networks (ANNs). An ANN is a collection of interconnected units or nodes, which loosely model the neurons of the brain.

Fuzzy logic and ANNs are two important and distinct strategies used in artificial intelligence (AI) research. Each has properties that make it suited to tackling certain types of problems and not others. Fuzzy logic systems, for instance, are good at reasoning with information that's imprecise and can show how they arrived at decisions, but they're not equipped to figure out the rules by which they made those decisions. Neural nets, on the other hand, are well suited to applications like pattern recognition and understanding natural language but are notoriously black-box-like in their nature: they offer few clues as to how they produced their output. Because of these limitations, some AI researchers are focused on creating hybrid systems in which the two techniques are combined. The hope is that by merging fuzzy logic and ANNs the result, known as fuzzy neural networks or neuro-fuzzy systems, will overcome the handicaps of the separate approaches and be effective in solving a variety of real-world problems.

The past hundred years or so has seen the rise of many other types of logic that differ from the classic or Aristotelian kind. Among these is intuitionistic logic, which stems from the belief that maths doesn't already exist, waiting to be discovered, but instead is constructed purely in the mind. This philosophy, whose leading advocate, in the first half of the twentieth century, was the Dutch mathematician and philosopher L. E. J. Brouwer, argues that mathematical proofs must be constructive. For instance, it would insist that if a proof included the statement 'there exists some X such that...', then, to be valid, the proof must include a specific example of an X that actually works. Intuitionistic logic was developed to the stage where it could be incorporated into mathematical logic, complete with symbols and formalised rules, by Arend Heyting, a student of Brouwer's at the University of Amsterdam. Notable among its features are the absence of the double negative (so that not-not-A doesn't imply A) and of the law of the excluded middle (meaning that A or not-A isn't necessarily true). Something else intuitionistic logic doesn't allow is proof by contradiction.

Suppose you have an umbrella. On Monday evening the umbrella is dry but on Thursday morning it's wet. The statement we're trying to prove is 'Either it rained on Tuesday or it rained on Wednesday.' Provided that the umbrella only gets wet when it rains (and dries out very, very slowly!), we can reach a solution using classical logic in the following way. Start by assuming that the conclusion is false: namely, that it rained on neither Tuesday nor Wednesday. If this were the case, then it follows that since the umbrella was dry on Monday evening it would still be dry on Thursday morning. But this leads to a contradiction since we know that on Thursday morning the umbrella is wet. Thus, our assumption that it rained on neither Tuesday nor Wednesday is false and it must be true that it rained on either Tuesday or Wednesday.

However, we've just given a proof by contradiction, which under intuitionistic logic is a no-go. In fact, intuitionistic logic won't even entertain the statement 'Either it rained on Tuesday or it rained on Wednesday.' The reason is that to prove a statement of the form 'A or B' you have to give either a proof of A or a proof of B; what you can't do is say 'either A or B is true but we don't know which.' In the case of the umbrella, you can't prove that it rained on Tuesday (maybe it was sunny on Tuesday and it rained on Wednesday) and you can't prove that it rained on Wednesday (maybe it was sunny on Wednesday and it rained on Tuesday) so you've no way of proving the seemingly obvious statement 'Either it rained on Tuesday or it rained on Wednesday.' You can prove that its negation – it rained on neither Tuesday nor Wednesday – is false, but that doesn't help as you then need to use the law of double-negation, which intuitionistic logic doesn't admit.

This simple example may suggest that intuitionistic logic, and constructivism on which it's based, have some serious shortcomings, at least when it comes to resolving everyday problems. But the reason constructivism came about, in the early part of the twentieth century, wasn't to address questions to do with wet or dry umbrellas. The issue then at hand was set theory. Specifically, there was a debate over whether to include in set theory the so-called axiom of choice (AC). In plain (and slightly imprecise) language, AC says that if you have a collection of buckets each containing at least one item, it's possible to select exactly one item from each bucket, even if the collection is infinitely large. Constructivists, like Brouwer, objected to AC because it doesn't specify *how* to choose one item from each bucket (or, to phrase it mathematically, one element from each set), and, in fact, it's often not possible to do this when an infinite number of buckets (or sets) is involved.

One of the consequences of AC is perhaps the most bizarre theorem in all of maths: the Banach–Tarski theorem. This states that a ball can be cut up and the pieces rearranged to form two balls, each the same size as the original, with no gaps or spaces. Sadly, the Banach–Tarski magic works only with mathematical balls and not the kind found in the real world, which are made of discrete atoms and molecules. Moreover, it only works in a world where constructivism and intuitionistic logic is denied. In the system of logic that Brouwer favoured, the Banach–Tarski theorem simply can't arise, nor can many other theorems that start from the assumption of the axiom of choice.

Nowadays, the general consensus is that non-constructive proof is just as valid as the constructive variety, and, although the former makes more assumptions, it comes with certain advantages. In 1928, David Hilbert wrote: 'Taking the principle of excluded middle from the mathematician would be the same, say, as proscribing the telescope to the astronomer or to the boxer the use of his fists.'

While some systems of logic are built around particular philosophies of mathematics, others are guided more by practical considerations. Boolean logic, for instance, arose when people became interested in using machines to do complicated calculations. In time it became the logic that guides the design and operation of electronic computers. But Boolean logic works only in the world of classical physics, where an object is always at a definite place in time and space and is in a clearly defined state. With the advent of quantum mechanics, it became clear that on an atomic and subatomic scale nature follows a different set of rules. There are limits to how accurately certain pairs of quantities, such as position and momentum, and energy and time, can be known. Also, the state of a particle is indeterminate unless and until a measurement is made to establish it.

In 1936, American mathematicians Garrett Birkhoff and John von Neumann wrote a paper in which they set out a new set of principles for dealing with the behaviour of things in this strange new realm and called it 'quantum logic'. It's the first example of a system of logic dictated entirely by breakthroughs, not in mathematics, but in physics – in the observed way that phenomena unfold on a very small scale. Whereas the algebra of Boolean logic is applied primarily to conventional computers, which work with binary digits, or bits, the main application of quantum logic is to an extraordinary new generation of computers. Quantum computers, working with quantum bits, or qubits, and guided by quantum logic, will soon be helping us design innovative medicines at the molecular level, optimise travel schedules for passengers and goods, and solve some problems in mathematics that are beyond the reach of even the fastest classical computer.

It may seem strange that there are so many different types of logic – Aristotelian, Boolean, fuzzy, quantum, and others that we haven't mentioned here. But, in essence, the differences aren't in the logic but in the thing being discussed: whether, for instance, it's a classical or a quantum system, or whether it can be answered using simply 'true' or 'false' or requires degrees of truth in between. Logic has been adapted to apply to very different situations such as the analysis of natural language or the development of new areas of mathematics with novel sets of axioms. It's sometimes asked whether elsewhere in the universe, and perhaps in other universes, mathematics and logic could be different. Almost certainly the answer is 'no': mathematics and logic are different nowhere, and true everywhere.

CHAPTER 12

Is Everything Mathematical?

> I think that modern physics has definitely decided in
> favour of Plato. In fact the smallest units of matter are
> not physical objects in the ordinary sense; they are forms,
> ideas which can be expressed unambiguously only in
> mathematical language.
>
> — Werner Heisenberg

DIG DEEP ENOUGH into anything, it seems, and you find
maths. That observation has led some mathematicians and
philosophers to the remarkable conclusion that everything,
including ourselves, is part of a mathematical structure. At
first, this seems incredible. After all, we live in a world full
of colours, emotions, sensations, and experiences, which we
can't (for now at least) transform into numbers or equations.
Surely the universe would be a sterile place – a ghostly shadow
of the real thing – if maths was all there was to reality. Yet,
in the end, all matter is made of fundamental particles, such
as electrons and quarks, the properties of which appear to
be purely mathematical.

As soon as we try to chase down an individual particle,
like an electron, it seems to lose its substance and wash out

into a wave of probability. What we'd taken to be physical –
hard-edged and tangible – melts into something abstract and
without substance. Space, too, on close examination, reduces
to a mere mathematical structure.

This idea that maths underpins the physical universe, and
is perhaps its foundational essence, has been hugely influential in Western thought. It was first expressed strongly in
the sixth century BCE by Pythagoras and his followers, who
lived by the motto, 'All is number.' To the Pythagoreans,
each number had a certain meaning and character. Odd
numbers were male, even were female; one was the number
of reason, two of opinion, three of harmony, four of justice, and so on. Such mystical beliefs may have been picked
up by Pythagoras during his travels in Egypt and Babylon,
regions where numerology was popular. But there's no
question that the Pythagoreans were also powerfully influenced by observations of the world around them, especially
their discoveries in music. They noticed that harmonious
notes produced by a vibrating string, stopped at different
lengths, could be characterised by simple numerical ratios.
Halving the length of the vibrating part of the string gave
a note that was in unison with, though an octave higher
than, the original. Stopping the string two-thirds the way
along gave rise to a perfect fifth. Not satisfied with a mere
Earthly connection between maths and music, Pythagoras
projected the link into the heavens. In a theory that became
known as the Harmony of the Spheres, he taught that
the Sun, Moon, and planets each give off a unique tone
as they move around their orbits. It was a belief that carried through into Renaissance times and, in particular, to
Johannes Kepler, who gave us the laws of planetary motion
and yet also argued passionately for this antiquated notion

of cosmic music in his *Harmonices Mundi* (The Harmony of the World).

Pythagoras and those who congregated around him were obsessive and over-the-top in their convictions, as cult members typically are. But there's no question that their deep fascination with numbers spurred the development of maths and had a crucial impact on future generations of philosophers, scientists, and mathematicians in the West. Their devotion to everything numerical led them to classify numbers as even and odd, prime and composite, perfect, and friendly. They introduced figurate numbers – triangular numbers, rectangular numbers, and the like – as a way of representing numbers geometrically. They discovered (and detested) irrational numbers and showed how to construct the five regular solids: cube, tetrahedron, octahedron, dodecahedron, and icosahedron. In their drive to demonstrate that maths was the key to understanding everything in existence, they pioneered number theory and set the scene for modern physical science, which assumes from the outset that it will find maths behind every phenomenon it investigates.

Aristotle, who lived a couple of hundred years after Pythagoras, didn't buy into the harmony of the spheres – he was too much of a materialist for that. But he certainly backed the notion that maths was fundamental. 'The principles of mathematics', he wrote, 'are the principles of all things.' The same sentiment has echoed down the ages, right up to the present day. Galileo, a scientist in the true sense that he experimented to test his theories, asserted: 'The book of nature is written in the language of mathematics.' French philosopher and mathematician René Descartes said: 'With me, everything turns into mathematics.' Many contemporary physicists think along much the same lines. In the words of string theorist and mathematician Brian Greene: 'Physicists have come to realise that mathematics, when used with sufficient care, is a proven pathway to truth.'

It's a common – in fact, conventional – stance among scientists and engineers that maths is the best and most precise way to explain the way the world works. There's no doubt of the effectiveness of formulae such as $E = mc^2$, Newton's laws of motion, and the equations of general relativity. As far as these and other familiar mathematical rules of nature go, the matter isn't in dispute. If you want to land a spacecraft safely on Mars, you need to do the maths – solve the equations that guide the vehicle's motion – otherwise it will never happen. Maths informs the design, and explains the results, of high-energy experiments by particle physicists, is essential to every successful large engineering project, and predicts the motion of all objects from cannonballs to comets.

Yet while these examples demonstrate how useful maths can be, do they imply that everything in the physical world follows mathematical rules and that maths is somehow the bedrock on which reality rests? It's very easy to be persuaded so. American theoretical physicist Steven Weinberg has written about how maths seems to have prior knowledge of what's going on in the universe: 'There is a spooky quality about the ability of mathematicians to get there ahead of physicists. It's as if when Neil Armstrong first landed on the Moon he found in the lunar dust the footsteps of Jules Verne.'

In a famous 1960 paper, 'The Unreasonable Effectiveness of Mathematics in the Physical Sciences', Hungarian-American theoretician Eugene Wigner noted how 'The mathematical formulation of the physicist's often crude experience leads in an uncanny number of cases to an amazingly accurate description of a large class of phenomena.' A well-known example of this 'unreasonable effectiveness' is Newton's law of gravitation. Newton saw a connection between the parabolic paths followed by projectiles on Earth and the motion of the Moon and planets in their elliptical orbits. At the time he put forward his law of gravity he could verify it to an accuracy of only about four percent. Today, we know it holds good to better than one ten thousandth of a percent. In his essay, Wigner concludes: 'While science is composed of laws which were originally based on a small, carefully selected set of observations, often not measured very accurately, [these] laws have later been found to apply over much wider ranges of observation and much more accurately than the original data justified.'

Physics is replete with instances of mathematical predictions running ahead of observational evidence. It's also true that areas of maths have been developed for which a perfect match was later found, completely unexpectedly, with aspects of the real world. This happened in the case of matrix algebra, the rules of which were first figured out by English mathematician Arthur Cayley in the mid-1850s. About seventy years later German physicist Werner Heisenberg, working in collaboration with Pascual Jordan and Max Born, realised that these rules for manipulating matrices were formally identical to methods he was using to understand how particles behave at the quantum level. Future applications of matrix mechanics, in situations beyond those that Heisenberg foresaw, allowed predictions to be made that agree with experimental data to within one part in ten million.

Time and again, scenarios that arise from the manipulation of symbols are played out exactly in the real world. As we saw in Chapter 6, the equations of general relativity insisted that the universe as a whole should be expanding (or contracting). Einstein didn't believe this, and so he invented his cosmological constant to counteract the expansion. Yet the equations turned out to be correct: at some level, they 'knew' about the growth of the cosmos before humans did. The Higgs boson, too, popped out of the maths forty-eight years before it was finally detected by an experiment at the Large Hadron Collider. Somehow, knowledge of the Higgs was contained in the equations that described it decades before it was manufactured and observed in physical reality. In the words of Brian Greene, 'Maybe it's because math *is* reality.'

English mathematical physicist Roger Penrose sees a grand self-consistent and self-sustaining loop emerging from this worldview:

> We have a closed circle of consistency here: the laws of physics produce complex systems, and these

The Large Hadron Collider.

> complex systems lead to consciousness, which then
> produces mathematics, which can then encode in a
> succinct and inspiring way the very underlying laws
> of physics that gave rise to it.

Some regard this neo-Pythagorean view of the relationship
between maths and reality as becoming more and more
compelling as physics, propelled by ever more abstruse and
abstract calculations, continues to make remarkable progress. We're now able to foretell with exceptional precision
much of the behaviour of matter and energy, from the scale
of subatomic particles to that of superclusters of galaxies.
If only, at some point, we could forge a new description of
gravity from a marriage of quantum mechanics and general
relativity, we might have in our grasp, so the hope goes, a
'theory of everything'.

It's impossible to deny the power of the equations that
describe how nature operates across the unimaginably broad
spectrum of lengths, times, masses, and energies that char-
acterises the universe. It seems inevitable, too, that physics
will become vastly more capable in the years ahead as the
maths it deploys offers a theoretical description of the world
in increasingly exquisite detail. Already it's clear that the
maths used in science is far more than just a handy notational
system: it's a highly effective way of modelling the universe.
The question is whether it runs deeper than that and, in the
final analysis, maths serves as a window on reality itself.

Swedish-American cosmologist Max Tegmark has taken
the ultimate and most extreme step in this direction of argument. He's returned full circle to undiluted Pythagoreanism
and its fundamental tenet: all is number. In what he calls
his 'mathematical universe hypothesis', or 'mathematical

monism', Tegmark denies the existence of *anything other* than mathematical objects. Somewhat disturbingly, this would include even ourselves and the contents of our minds and awareness. 'Consciousness', says Tegmark, 'is probably the way information feels when it's being processed in certain, very complicated ways.' If he's right then the same kind of unification that's already taken place with regard to electricity and magnetism, matter and energy, and space and time, would, at some point, see mind and maths brought together. If that ever happened, our experience of being, and of being someone in the world, would come to be regarded as just another manifestation of data being shuffled around inside some all-encompassing cosmic computer. The realisation that everything that makes us who we are could be reduced to maths and nothing more would represent the final vindication of the core belief held by Pythagoras and his followers two and a half thousand years ago.

But there's another side to the debate about maths and its status in the world. According to this alternative outlook, maths is no more than a product of the human intellect – a mere tool that we use to describe and explain certain aspects of nature. The distinction between this and what the new Pythagoreans have to say is important and more than just a philosophical issue. If maths does prove to be just a mental construct, then, at some point, its limitations will become apparent and we'll have to accept that there are constraints on how much it can tell us about reality.

Without becoming a card-carrying Pythagorean, it's become common for scientists to remark on the power of maths to describe, with surprising precision, the world 'out there'. Einstein asked: 'How can it be that mathematics, being after all a product of human thought which is independent of experience, is so admirably appropriate to the objects of reality?'

But are we kidding ourselves that maths is always so good at modelling the way things actually are? Looking at school physics and maths problems, you can see how massively simplified they have to be in order that students can figure out exact solutions to them. Expressions such as 'frictionless surface', 'perfectly elastic collision', and 'particles joined by an inextensible string', are common. In the same way, in pure maths, only certain types of equations, integrals, and so on, are ever tackled. Partly, of course, this is because young people haven't yet learned all of the advanced techniques available for solving more complicated problems. But that isn't the main issue. The fact is, it doesn't matter how many years you study maths and physics, as almost no situation in the real world can be modelled precisely by mathematics, and many phenomena can't yet be described mathematically at all except perhaps in the broadest terms.

To give an example, say a leaf is floating down a steadily flowing river. It's easy to work out how far the leaf will travel in half a minute if we know how fast it's moving. But in saying this we've already isolated and simplified a tiny facet of a real situation that we know will be mathematically tractable. How about if we ask for the exact motion of the leaf over the course of that half-minute? This would involve being able to predict *in advance* how the water around the leaf would flow, millisecond by millisecond, including small disturbances caused by currents, bends in the river, unevenness of the riverbed, and the action of any wind at the surface of the water. Or take another example. Say you drop a stone from a height of two metres. It's very simple to work out how long it will take to hit the ground and the speed at which it will be travelling when it lands, and your calculations will match pretty well any measurements you make of the drop. But try to model the exact motion of a feather as it descends, out in the open with a gentle breeze blowing, from the same height, and you'll have no luck at all.

Almost every activity in the physical universe is messy. It involves complicated objects, forces pulling this way and that, and incredibly intricate dances of numerous components, such as molecules, raindrops, or stars. One of the reasons we get the impression that mathematics is successful in describing the real world is that we cherry-pick problems for which we've found a way to apply the maths and laws that we know. Having said this, there's no denying how astonishingly useful maths is in situations where it *can* be applied effectively. No bridge, tunnel, dam, or skyscraper is built these days until a thorough analysis has been carried out that involves working out stresses and strains on every part of the structure under a variety of conditions. Maths is the indispensable tool of the cosmologist, the theoretical physicist, the spacecraft engineer, and the meteorologist. It's exceptionally important to us in predicting and explaining how certain things or systems behave. But still, the fact remains, we tend to focus on where it's been successful and ignore the countless examples in nature in which, for the time being at least, it's almost entirely ineffective.

Some problems that can't be solved exactly, or by analytical methods, are at least open to approximate solution by numerical methods. The advent of powerful computers has allowed scientists and mathematicians to simulate systems that are otherwise too complex to deal with. Weather forecasting's a case in point. In the past, knowledge about whether it would rain tomorrow or be sunny was largely in the lap of the gods.

Local experts, based on long experience and old records, might do better than the average person at predicting meteorological conditions over the next day or so, but flipping a coin might be almost as effective. Today, supercomputers can crunch through elaborate systems of equations, transformed into digital form, feeding in data supplied by satellites and ground-based weather stations, and give useful forecasts out as far as ten days or so. That's not only handy to us in planning day trips to the beach, but it's vital for shipping and air transport, and can save lives if a hurricane is bearing down on a population centre. Nevertheless, such forecasts are often still inaccurate in detail and fail altogether beyond about a week and a half in advance.

In years to come, we'll have more powerful computers, enhanced by artificial intelligence, fed by more detailed data. But at some point, the complexity of almost all natural phenomena, whether it concerns the weather or the evolution of a galaxy, will overwhelm our ability to simulate it. The universe itself is, ultimately, the only accurate, real-time simulation of itself. Its contents and behaviour are inherently too noisy and intricate to allow us to create a shorter, idealised version of the real thing. In terms of information, physical reality is 'incompressible' – in most situations, we can't usefully apply maths to create a more compact simulation or solution. The reason so many scientists have expressed agreement with Eugene Wigner's claim of the 'unreasonable effectiveness' of maths is that they focus on where maths has been successful in modelling outcomes. There's a tendency to overlook where, in the vast majority of cases, it fails to deliver the same degree of effectiveness and elegant compression.

It's also worthwhile thinking about the assumptions we tend to make about maths, even at the most basic level.

Counting things, for instance, isn't as straightforward as we often assume. We count items that belong to categories: cats, pebbles, stars, and so on. But each of these items – a cat, for instance – is a collection of things in its own right. A cat is a complex bundle of stuff, all the way down to the molecular, atomic, and subatomic scale. There's no acknowledgement or allowance made for this mind-numbing complexity when we simply count a cat as 'one'. Furthermore, every cat is different, in size, colour, temperament, age, and a host of other factors. When we count 'five' cats, what exactly does that mean? Do we include wild cats as well as the domestic variety, and how about lions and tigers, and dead cats? There are major assumptions we make, right from the start, when counting things in the real world, which tend to be glossed over when considering the effectiveness of basic arithmetic. Effective it undoubtedly is, otherwise we'd never have developed and used it in the first place for practical, everyday purposes such as bartering, commerce, and keeping records. But it's as well to remember that number is an abstracted concept – an outcome of our labelling, categorising minds – not something that's necessarily inherent to the world outside.

Counting came about because it's a useful convenience. Keeping a tally of your income and expenses helps you be a successful merchant. For such purposes, you don't need to be philosophical about what you're doing. However, when it comes to asking about the basic nature of maths and its relationship to physical reality, details matter. When we count objects, do we know what those objects really are, and where they end and begin? Visually it may seem obvious, but our brains and senses have evolved to see the world in a particular way that helps our survival. The maths we've developed, at least in its original and most elementary form, was tailored to suit our immediate needs and way of life. Suppose creatures evolved on another world that weren't solid but cloud-like, or that they took the form of a living sea. To such amorphous beings, counting discrete objects might not seem so obvious or natural. How much of our maths then, from the simple concept of number to the most arcane theory at the frontiers of twenty-first-century mathematical research, is an artefact of the human condition? Those who oppose the Pythagorean notion that 'all is number' insist that there's no guarantee our mathematical descriptions are universally applicable. As useful as maths undoubtedly is, it may be far more limited in scope and power than is generally recognised.

There's also an elephant in the room whenever the topic of maths and reality comes up. The universe consists not just of inanimate objects engaged in some complicated dance that, if only we were clever enough, we'd see was choreographed by a master series of equations. It contains consciousness. Specifically, it contains *us* – assemblages of flesh and blood that experience 'what it is like to be' in the world. It might be going too far to say that consciousness is an embarrassment to science, but ever since Galileo there's been a sustained effort to downplay its importance. To a large extent it's regarded as an epiphenomenon, an almost superfluous effect due to the brain's other workings – like mist hanging over a lake.

This tendency of science to overlook the one thing that's most important to us as humans – our awareness – is no acci-dent. It was Galileo, in the seventeenth century, at the dawn of modern physics, who argued that reality is divided into two types of quality: that which can be measured and that which can be experienced. Measurable qualities he referred to as being 'primary'. These include mass, size, temperature, location, and other aspects of things that, by virtue of being measurable in some way, can be expressed in mathematical terms. 'Secondary' qualities, on the other hand, exist only in the minds of sentient observers and thus have no place of importance in the material world. Into this category of phenomena, which slip through the clutches of mathematics, are colours, sounds, and other sensations, along with all emotions and feelings such as pleasure and pain. It's true that there are measurable, physical correlates to each of the components of our inner world – wavelength and radiance, for example, to our experience of colour and brightness – but science is equipped to handle only the former. Physics can deal with wavelengths of the electromagnetic spectrum because it can hold up an instrument to measure these wavelengths and convert them into numbers and then work on them with equations. Physicists are happy to talk about wavelengths in the range 700 to 635 nanometres and comfortable even to describe these as falling within the red part of the spectrum. But ask them to talk about the *quality* of redness and they fall silent. To be fair, it isn't just a problem with physics. It's impossible by means of any symbolic or linguistic device, or any purely intellectual means, to convey the sensation of redness unless you've actually experienced it. A person who's been blind since birth, or who lacks colour vision, can never know what redness is. They may understand every aspect of the maths of light and the science of electromagnetic waves, but the sensory correlate itself will always be missing from their picture.

It's obvious why modern science is mainly interested in measurable stuff. If it can't, in the end, analyse the data it collects, it has nowhere to go, and it can only analyse what's been measured. Paradoxically, then, the greatest strength of science is also its greatest weakness. It excludes what it can't measure, and thereby turn into numbers, so that it can unleash the power of mathematics. But, by largely excluding qualities that are dismissed as 'secondary', it fails to deal adequately with everything that's of greatest importance to us as living, breathing individuals.

Not all physicists are comfortable with the exclusion of qualities. Particle physicist and Nobel Prize winner Richard Feynman said in one of his lectures: 'The next great awakening of human intellect may well produce a method of understanding the qualitative content of equations.' But that's probably a pipe dream. The fact is that the qualia we experience in our minds and through our senses are not at some future stage going to emerge from our equations or our mathematics. They can't do that for a very simple reason: they've been intentionally omitted from the outset. No matter how much we improve our quantitative account of the world it will never be able to conjure up the secondary qualities that each of us, individually, regard as being primary. Would you exchange the sensation of colour or the feeling of love for the most fantastically detailed mathematical description of how these things come about?

Mathematics and physics appear so powerful because they avoid addressing those aspects of the world in which they're inherently weak or entirely powerless. Bertrand Russell, in *An Outline of Philosophy* (1927), put it this way: 'Physics is mathematical not because we know so much about the world but because we know so little; it is only its mathematical properties that we can discover.' And yet, having acknowledged that maths and its application to the physical universe – physics – will always appear aloof and almost irrelevant to our personal experiences, we're left still with that 'unreasonable effectiveness'. Maths works. Physics, which is based on maths, works. They enable us to do things – amazing things through technology – that would be impossible if we relied simply on our sensations and inner feelings. We know about dark energy and Higgs bosons, different types of infinity, and the maths of higher dimensional spaces, not because of the 'secondary' qualities that make life worthwhile but because we've learned how to separate the qualitative from the quantitative. The undeniable truth is that the world is more comprehensible to us now, through the application of maths and science, than it was to our ancestors, hundreds or thousands of years ago.

We've yet to fully understand the ultimate role that maths plays in the reality around us – and within us. Matter dances to the tune of mathematics. Mind perceives the existence of matter and explains its behaviour through mathematics. Without actuality, which requires matter and mathematics, there'd be no mind. Somehow, it seems, mind, matter, and mathematics rely on the presence of each other – essential elements in a self-sustaining and self-actualising cosmic triangle.

CHAPTER 13

The Next Fifty Years

When the time is ripe for certain things, these things appear in different places in the manner of violets coming to light in early spring.

– Farkas Bolyai

PREDICTING THE FUTURE is a risky business, almost guaranteed to make a fool of anyone who tries it. Eminent mathematical physicist William Thomson (Lord Kelvin) said in 1895: 'Heavier-than-air flying machines are impossible.' Only eight years passed before the Wright brothers proved him wrong, although even Wilbur Wright hadn't been optimistic about their prospects: 'In 1901, I said to my brother Orville that man would not fly for 50 years. Ever since I have... avoided predictions.'

Speaking in 1946, Darryl Zanuck, movie executive at 20th Century Fox, was scornful of the possibilities for television: '[It] won't be able to hold on to any market it captures after the first six months. People will soon get tired of staring at a plywood box every night.' Similar pessimism was expressed by some pundits about the prospects for the computer. In 1943, president of IBM Thomas J. Watson was of the view that 'there is a world market for maybe five computers.' In 1977, Ken Olsen, founder of Digital Equipment, announced: 'There is no reason anyone would want a computer in their home.' Within months, the hugely successful Apple II, TRS-80, and PET 2001 were launched and the home computer revolution had begun.

On the other hand, predictions have sometimes been overly bullish. The world is still waiting for the nuclear-powered vacuum cleaner, which, according to Alex Lewyt, president of vacuum cleaner company Lewyt Corp., in 1955, 'will probably be a reality in 10 years'. Even Arthur C. Clarke, who correctly forecast the importance of communication satellites, more than a quarter of a century before the first one was launched, was too eager in his prognostications about human space travel. Impressive though the International Space Station is, it pales next to the giant, wheel-shaped orbiting hotel and lunar transfer hub portrayed in *2001: A Space Odyssey*.

Predicting what's to come in mathematics is just as hazardous as in science and technology. Nevertheless, speculation is always fun and there are certain developments taking place in maths today that provide a basis for educated guesswork. Suppose that we could step into a time machine and be whisked forward fifty years. How would maths have changed? What would have been the big breakthroughs?

Some advances come out of the blue and take everyone by surprise. When Andrew Wiles announced in 1993 that he'd proved Fermat's last theorem, it came as a complete shock to most mathematicians. Even more unexpected was the discovery of quasicrystals – ordered structures that are realisations of the aperiodic tilings discovered by Roger Penrose a few years earlier and which many chemists had thought were impossible in nature. There'll always be such

A zinc-magnesium-holmium quasicrystal in the form of
a pentagonal dodecahedron.

surprises when exploring the unknown. A couple of trends
in maths, however, have become obvious over the past couple
of decades and seem bound to continue in the years ahead.

The first such trend is the rise of almost impenetrable
proofs – proofs that are so long and difficult that no one,
apart from their authors and a handful of other experts, can
understand them or has the ability to check them. In 2012,
for instance, Japanese mathematician Shinichi Mochizuki
claimed to have found a proof of the so-called *abc* conjecture. He published his work online in a series of four papers,
stretching in total to 500 pages of dense text and formulae.
As well as being dauntingly long, it involved an entirely new
type of maths with a name that sounds as if it were hatched
by a *Star Trek* scriptwriter: inter-universal Teichmüller theory
(IUT).

The *abc* conjecture is also known as the Oesterlé–Masser
conjecture, after French mathematician Joseph Oesterlé and

British mathematician David Masser, who formulated it in the 1980s. It's an important open question in number theory, which starts with the simple, innocuous equation $a + b = c$, where a, b, and c are coprime – in other words, numbers that don't share a common factor other than 1. It then asks about the 'radical' of abc, rad(abc). This is defined as the product of all prime factors that divide abc, ignoring higher powers; so, for example, rad(16) = 2. In most cases, it turns out that rad(abc) > c, but there are exceptions. In fact, there are infinitely many exceptions. What the abc conjecture claims, however, is that, if rad(abc) is replaced by any power greater than 1, the number of exceptions becomes finite. For instance, there are only finitely many triples a, b, c for which $c \geq$ rad(abc)$^{1.001}$.

If the abc conjecture could be shown to be correct, it would imply the immediate proof of at least sixteen other conjectures, including the Beal conjecture that we looked at in Chapter 9. These consequential, bonus results in a sense elevate the abc conjecture above Fermat's last theorem, which, despite being a lot better known, is, as far as we know, an isolated curiosity that leads on to virtually nothing else.

The trouble is that so few people have a thorough understanding of the subject, known as anabelian geometry, that Mochizuki's purported proof involves. As far as IUT is concerned, the number of experts dwindles to just a single individual – Mochizuki himself. Very likely his claimed proof wouldn't have received much attention at all had it not been for two factors: Mochizuki is a respected mathematician who'd already developed several important new theories in this area, and there's a general suspicion that the abc conjecture will turn out to be correct.

Several conferences were held in 2015 and 2016 to try to penetrate inter-universal Teichmüller theory and assess whether Mochizuki was on the right track. More than a hundred mathematicians took part but they faced a monumental task, as described by one of the participants, number theoretician and Stanford professor Brian Conrad:

> Very quickly, the experts realised that the evaluation of the work was going to present exceptional difficulties. The manner in which the papers culminating in the main result have been written, including a tremendous amount of unfamiliar terminology and notation and rapid-fire definitions without supporting examples nearby in the text, has made it very hard for many with extensive background in arithmetic geometry to get a sense of progress when trying to work through the material.

For the moment, the debate over whether Mochizuki's proof is valid or not continues to rumble on. The consensus seems to be that the proof falls short of what's needed and that there are specific weaknesses within it that won't be easy to fix. But, as Conrad's comments above make clear, resolving the issue isn't just a case of checking through the proof, line by line, and at the end, reaching a 'right' or 'wrong' decision. The language and concepts it uses are so unfamiliar and specialised that it's hard for mathematicians to grasp even the strategy behind the proof never mind its fine detail. This kind of situation, where claimed proofs involve very lengthy, highly technical, and novel terminology and ideas, is likely to become increasingly common in the future as mathematicians become more and more specialised in their fields of expertise.

Another significant trend in maths is the rise of computer proofs. The first of these, in 1976, finally laid to rest a problem that had taxed mathematicians for more than a century: the four-colour theorem. This states that it's possible to colour any map on a plane using just four colours so that any two regions that share a common border are coloured differently. It was first proposed in 1852 by South African Francis Guthrie, whose younger brother, Frederick, was at the time a student of Augustus De Morgan at University College London. Frederick passed along his brother's theory to De Morgan who, intrigued, immediately wrote to his friend William Rowan Hamilton (pioneer of quaternions) in Dublin:

> A student of mine asked me today to give him a reason for a fact which I did not know was a fact – and do not yet. He says that if a figure be anyhow divided and the compartments differently coloured so that figures with any portion of common boundary line are differently coloured – four colours may be wanted, but not more – the following is the case in which four colours are wanted. Query cannot a necessity for five or more be invented... If you retort with some very simple case which makes me out a stupid animal, I think I must do as the Sphinx did...

A reply arrived from Hamilton three days later:

> I am not likely to attempt your quaternion of colour very soon.

Other mathematicians, on learning of Guthrie's theory, quickly took an interest in it. Eccentric American philosopher

An extract from De Morgan's original letter to Hamilton
about the four-colour conjecture, 23 October 1852.

and logician Charles Peirce wrote on the four-colour theory
in the 1860s and retained a fascination for the problem
throughout his life. In 1878, English mathematician Arthur
Cayley put a question to the London Mathematical Society
asking if the problem had yet been solved. The following year
it seemed it had. London barrister Alfred Kempe announced
in the journal *Nature* what he claimed was a proof. Kempe
had studied maths at Trinity College, Cambridge, where
Cayley was one of his teachers. At Cayley's suggestion Kempe
submitted his proof to the *American Journal of Mathematics*
and it was published in 1879.

Kempe used a concept in graph theory known as a planar graph. A graph is just a mathematical structure in which points (also called nodes or vertices) are connected by lines (also called edges). A graph is said to be planar if it can be drawn on the plane with no two edges intersecting. Kempe's central idea was to show that there's a certain set of configurations that must appear in all planar graphs and then show that, if these graphs need five colours, then so does a smaller graph. He found a very small unavoidable set – namely that of a triangle, a square, or a pentagon. However, as it was later discovered, he'd failed to prove that if a graph with a pentagon needs five colours, so does a smaller graph.

Kempe won plenty of praise for his work and the ingenuity of a device he brought to bear in it, which came to be known as Kempe chains. His contribution to solving one of the great problems in topology earned him a fellowship of the Royal Society and he also served as president of the London Mathematical Society. But eleven years after his proof was published it was found to contain the slight but fatal oversight just mentioned. The flaw was spotted by Durham University lecturer Percy Heawood, who was described in a 1963 article in the *Journal of the London Mathematical Society* as 'an extravagantly unusual man'. The article went on:

> He had an immense moustache and a meagre, slightly stooping figure. He usually wore an Inverness cape of strange pattern and manifest antiquity, and carried an ancient handbag. His walk was delicate and hasty, and he was often accompanied by a dog, which was admitted to his lectures…

Although Heawood returned the four-colour theorem to the shelf of 'open' problems, he was able to use Kempe's strategy to prove the five-colour version of the theorem and generalise the four-colour conjecture beyond the plane to other types of surface. Only in 1976 was the original four-colour theory finally shown to be true once and for all – but not solely by human hand.

Decades of effort by mathematicians produced many related breakthroughs and culminated in the final solution to the four-colour problem itself. Kenneth Appel and Wolfgang Haken of the University of Illinois at Urbana-Champaign reduced the task to testing 1,936 different map configurations, involving 10 billion individual cases, to see if the minimum four-colour criterion was satisfied by each. No team of human calculators could have done the job in anything like a reasonable amount of time, so Appel and Haken wrote a program to do it by computer. In the mid-1970s, the first supercomputers were being built. Appel and Haken wanted to use one of these – the world's first massively parallel computer, the ILLIAC IV, which was taking shape on their doorstep at the University of Illinois – to crunch the problem. However, after being told it wasn't yet available, they found a worthy substitute in a Control Data 6600 at Brookhaven National Laboratory – a machine that ran through all the necessary calculations in about 1,200 hours.

The results were checked using different computers running different software. In the end, the proof stood: no map exists for which more than four colours are needed to avoid any two neighbouring regions being coloured the same. But a proof achieved with computer aid? There was an immediate outcry from the maths community. Some mathematicians and philosophers argued that only proofs obtained by humans could be considered legitimate. Among the reasons given were that machines lacked creativity and could only muscle their way to a result in a manner that was inelegant and amounted to mere symbol manipulation. Others were concerned about the reliability of computers and of the programs and algorithms they ran. However, people too can and do make mistakes, so this isn't a weakness exclusive to machines.

Like it or not, the proof of the four-colour theorem marked a watershed in mathematics. With it came the realisation that not only were there proofs so complicated that no human, or human team, could check them, but also that a potent tool was now available to do this instead. It's since become increasingly common to enlist high-speed computers in combing through all possibilities of complex problems to verify proofs by brute force. Objections have been raised to computer-assisted proofs on the grounds that they can't be validated first-hand by people. However, programs have been developed, such as Coq (named after one of its inventors, Thierry Coquand), that function as proof assistants and produce results in a form that can be surveyed and examined by human checkers.

The fact is that computers have already become an indispensable partner in some areas of mathematical research. Inevitably, their influence and involvement are going to expand in the years ahead. A major threshold will be crossed when computers are able to come up with their own proofs. Developments in artificial intelligence suggest that this isn't some far-off pipe dream. In 2016, the program AlphaGo, developed by Google, defeated Lee Sedol, one of the world's best Go players, in a five-game match. In 2017, AlphaGo was surpassed by another Google program, AlphaZero, which can also play chess and the Japanese board game shogi. Both AlphaGo and AlphaZero use a very different approach to that of traditional chess computers such as Deep Blue, which defeated then-World Champion Garry Kasparov in 1997. While Deep Blue's advantage lay in being able to check vast numbers of possible moves, AlphaZero can recognise abstract patterns – a skill which is essential to Go and the reason why many top Go players were sceptical that computers could surpass the best human players any time soon. It's this type of artificial intelligence, one that can recognise abstract connections and ideas, which will be necessary if computers are ever to become superhuman at mathematics itself. As computers rival and exceed the abilities of their flesh-and-blood creators in such skills, it may be that the proofs they generate will increasingly come to be seen as involving genuine creativity, just as human proofs are.

Two areas, in particular, may flourish as computers take on an increasingly greater role in mathematics. Experimental maths involves the use of computers to generate enormous data sets and then to search these for patterns that may form the basis of new conjectures and theories. Semi-rigorous maths is more controversial and anathema to the way mathematicians have traditionally gone about their work. It's been suggested that a time will come when decisions have to be made about the monetary cost, in terms of computer time and resources, of trying to prove (or refute) certain theorems. In the opinion of Israeli mathematician Doron Zeilberger situations may arise where it's possible to show that a definitive proof exists but that to obtain it would cost more than we're willing to pay. In these cases a semi-rigorous proof that offers near-certainty may be considered acceptable. He envisages the abstract of a future paper that might read, for example: 'We show, in a certain precise sense, that the Goldbach conjecture is true with probability larger than

0.99999, and that its complete truth could be determined with a budget of $10B.'

The increasing use of computers in maths, especially for analysing vast amounts of data, also raises ethical questions. We're familiar with the relevance of ethics in biological – especially medical – research and even physics if it's being applied, for instance, in the development of weapons. Maths, on the other hand, seems to be so abstract as to be immune from considerations of right and wrong. That perception, however, is misleading. There are cases of people using mathematics without realising the ethical implications and also of misusing maths deliberately.

Cambridge Analytica, a company founded in 2013, employed data mining and analysis to manipulate election results in countries around the world, most notably that of the 2016 US presidential election in favour of Donald Trump's campaign. However, its ability to do so rested on the way Facebook handled data obtained from its users. Like many companies, Facebook collects as much customer data as possible and then looks at it very closely to harvest specific, and commercially valuable, information. It's relatively easy, for instance, to determine users' locations and, from these, to draw conclusions about the activities in which they're engaged. It's also possible to use the data collected to calibrate an individual's beliefs – to project which way they're likely to vote in an election and also to gauge the strength of their political views. A key reason for doing this kind of analysis is to enable advertisers to target their ads at people who are more likely to be receptive to them. The process can also lead to feedback loops, as anyone who uses YouTube will be aware.

Feedback loops often occur when the output of algorithms is reinforced based on human decisions. If you watch a

YouTube video with a certain political viewpoint, YouTube's algorithms will recognise this and recommend more videos with similar content. People have a tendency to click on these recommendations, which are displayed prominently at the top of the screen, and, especially with the Autoplay feature enabled, to simply watch whatever is played next. The result is a feedback loop, wherein watching one video with a certain political leaning leads people to watch more and more that push the same message, but with the viewpoints expressed becoming more extreme over time. Once YouTube has identified a group of people who watch a certain type of video, it then allows advertisers to target their ads at this audience, thus monetising it and providing greater incentive to attract more people to the same kind of content.

In the case of Cambridge Analytica, it first developed an app called 'This Is Your Digital Life'. The app harvested data from all its users and their Facebook friends, and determined their political leanings precisely enough to deliver ads deemed most effective at influencing the election results. For instance, if someone were determined to be a Hillary Clinton supporter, they might receive an ad about how well Clinton was already doing, which might make them less likely to think that they needed to vote when election day came. On the other hand, a moderate Republican who otherwise might not have bothered to vote might be fed an ad about how important voting is, without mentioning any particular party or candidate – an unnecessary detail given that Cambridge Analytica had already determined that if they did vote they'd be likely to vote for the Republican candidate, Donald Trump. The cleverness of this system, and what makes it so subversive, is that it's difficult to trace. Each user sees at most one advert and they seem to come from completely different sources.

Someone who received a commercial about the success of Clinton, for instance, would probably assume it came from the Democratic Party, whereas someone who received an ad encouraging them to vote would probably assume that the source was an independent, non-partisan election commission. By concealing its tracks, Cambridge Analytica was able to influence subtly a number of elections around the world, including the outcome of the 2016 US presidential election. Not until 2018 were the company's nefarious activities brought to light and it was forced to shut down. Facebook was fined £500,000 for failing to safeguard users' personal data. However, similar ethical violations, engineered at both a corporate and state level, are doubtless ongoing and will be a feature of the online environment for many years to come.

The collection and usage of data can also bring unexpected risks. Strava, a social fitness network and app, allows users to track their exercise in detail and, with an optional accessory, can measure more detailed information such as heart rate. The Strava company later decided that it would release all of its data online, thinking it would be safe to do so because everything was anonymised. But someone with access to the data can easily trace each route that's been jogged back to a particular address, which could then allow the individual who uses that route to be identified. What's more, if it's known that this person goes jogging at, say, 8:20 am every day, their whereabouts can be more or less precisely tracked. From this it's clear that the idea of trying to anonymise data by simply removing names is hopelessly naïve.

A potentially more serious problem stemming from Strava's collection of joggers' information was spotted in 2018 by Nathan Ruser, an Australian university student studying international security. The data came, of all places, from a region of Afghanistan that was largely desert. The area was occupied mostly by nomadic goat herders – not the obvious demographic for a fitness app. Yet Ruser noticed that the Strava data were showing many joggers in the area who were travelling along paths that were almost perfect squares. Frequently, the paths would enter the squares and move about inside them. These were no Afghan goat herders, Ruser realised: they were American soldiers jogging around and inside secret US military bases. In its enthusiastic collection of data, Strava had inadvertently revealed not only the location of the military bases but also detailed maps of their interiors.

Such leakage of sensitive data might come as negative or positive news, depending on who you are. For instance, if you worked for the US Department of Defense, you'd probably be outraged at the potential threat to national security from the location of bases being revealed for all to see. On the other hand, if you were an agent of the Taliban seeking a military advantage, or an employee of Amnesty International wanting to know if the US was wrongfully arresting and detaining people, you might regard the Strava information as being very helpful. In any event, incidents such as those involving Strava and Facebook show that if you keep vast quantities of data, there'll always be the risk of some of it being used for purposes you didn't expect and which may cause harm.

Speculating on the future of mathematics is nothing new. It's been an occupation of eminent mathematicians for well over a century. According to Henri Poincaré, writing in 1908:

> The true method of forecasting the future of mathematics lies in the study of its history and its present state.

It's useful, in this context, to look at what mathematicians in the past have regarded as being the greatest challenges that faced their subject, and comparing these with what's been achieved since. At the start of the twentieth century, David Hilbert, probably the most influential mathematician of his time, published a list of twenty-three problems that he believed were the most important and exciting ones that remained to be solved. In a famous address he gave to the Second International Congress of Mathematicians in Paris in 1900, he said:

> The great importance of definite problems for the progress of mathematical science in general…is undeniable…[for] as long as a branch of knowledge supplies a surplus of such problems, it maintains its vitality…

Many of Hilbert's problems were solved in subsequent decades, and on each occasion it marked an important step forward for the subject. Some have been reassessed as being too vague or ill-defined to allow a definite solution, while one is now considered to be a problem in physics rather than maths. Only three remain open, including the most famous and, by general consensus, the most important unsolved problem in mathematics – the Riemann hypothesis, which deals with the distribution of prime numbers.

The nearest equivalent today of Hilbert's twenty-three problems is the seven Millennium Prize Problems endowed by the Clay Mathematics Institute with a \$1 million reward. Only one of these twenty-first-century challenges has so far been answered to the satisfaction of the maths community – the Poincaré conjecture – by reclusive Russian mathematician

Grigori Perelman, who refused to accept either the million-dollar prize or the Fields Medal that was offered to him.

Of tremendous practical interest is another of the Millennium problems, known as P versus NP (discussed in detail in Chapter 5 of *Weird Maths*). This amounts to finding out whether or not, in all cases, if the solution to a problem can be verified quickly, it can also be found quickly as well. If it turns out that $P = NP$, then all questions in maths for which a purported answer can be verified quickly can also be solved quickly. The consequences would be immense: scheduling transportation could be done optimally, enabling the most efficient possible movement of people and goods; productivity in manufacturing could be sped up while creating less waste; and all kinds of complex scientific simulations, such as that of protein folding, would be rendered tenable. American computer scientist Scott Aaronson put it this way: 'If $P=NP$, then the world would be a profoundly different place than we usually assume it to be. There would be no special value in "creative leaps", no fundamental gap between solving a problem and recognising the solution once it's found.' On the flip side, if it turns out that P doesn't equal NP, it would mean that some problems would take an astronomical amount of time to solve, no matter what amount of resources, data, or expertise we threw at them.

Mathematicians have made good progress in recent years by transferring tools from one field and applying them to problems in another. Often they've done this by translating the terminology of an unsolved problem, say in number theory, into the language of a different subfield, such as topology. Breakthroughs achieved in this way suggest that perhaps what appear to be very different parts of maths are, in fact, intimately connected. Some have even gone so far as to wonder if a mathematical theory of everything lies within our grasp – the equivalent of the all-embracing theory that scientists have speculated may link together the most fundamental aspects of the universe.

There's never been a more exciting time to be involved in mathematics. Just as astronomers look out on a cosmos full of unexplored worlds and exotica such as black holes and dark energy, mathematicians marvel at the unsolved mysteries of prime numbers and multidimensional geometries. They're beginning to discern a web of connections between seemingly disparate areas of maths through instruments such as homotopy theory and category theory. The use of computers in maths is on the rise and will accelerate as artificial intelligence becomes involved in exploring virgin tracts of the subject and pushing ahead at speeds far beyond any attainable by the human brain alone. We're at the dawn of a golden age of mathematical exploration, at the brink of glimpsing new vistas in number, shape, and symmetry, both wonderful and weird.

Acknowledgements

ONCE MORE WE'D like to thank our families for their love and support during the writing not only of this book but also of the entire *Weird Maths* trilogy. David is also grateful for the help of his old university friend Andrew Barker, for reading through the chapters and making many valuable suggestions. Last but not least, we'd again like to thank our editor, Sam Carter, and assistant editor, Jonathan Bentley-Smith, along with the rest of the wonderful staff at Oneworld who've made this series possible.

For more

WEIRD MATHS

look no further than

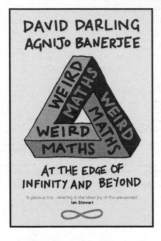

In the first instalment of the 'Weird Maths' series, David Darling and Agnijo Banerjee draw connections between the cutting edge of modern maths and life itself, on a quest to consider the existence of free will, how to see in 4D and the future of quantum computers. Packed with puzzles and paradoxes, this is for anyone who wants life's questions answered – even those you never thought to ask.

In the second instalment of the series, David Darling and Agnijo Banerjee go in search of the perfect labyrinth, journey back to the second century in pursuit of 'bubble maths', reveal the weirdest mathematicians in history and transform the bewildering into the beautiful, delighting us once again.

For even more weird maths, please visit weirdmaths.com